W9-AHV-181

Pesticides in Aquatic Environments

Environmental Science Research

Volume 1 – INDICATORS OF ENVIRONMENTAL QUALITY
Edited by William A. Thomas

Volume 2 – POLLUTION: ENGINEERING AND SCIENTIFIC SOLUTIONS
Edited by Euval S. Barrekette

Volume 3 – ENVIRONMENTAL POLLUTION BY PESTICIDES
Edited by C. A. Edwards

Volume 4 – MASS SPECTROMETRY AND NMR SPECTROSCOPY IN PESTICIDE CHEMISTRY
Edited by Rizwanul Haque and Francis J. Biros

Volume 5 – BEHAVIORAL TOXICOLOGY
Edited by Bernard Weiss and Victor G. Laties

Volume 6 – ENVIRONMENTAL DYNAMICS OF PESTICIDES
Edited by Rizwanul Haque and V. H. Freed

Volume 7 – ECOLOGICAL TOXICOLOGY RESEARCH: Effects of Heavy Metal and Organohalogen Compounds
Edited by A. D. McIntyre and C. F. Mills

Volume 8 – HARVESTING POLLUTED WATERS: Waste Heat and Nutrient-Loaded Effluents in the Aquaculture
Edited by O. Devik

Volume 9 – PERCEIVING ENVIRONMENTAL QUALITY: Research and Applications
Edited by Kenneth H. Craik and Ervin H. Zube

Volume 10 – PESTICIDES IN AQUATIC ENVIRONMENTS
Edited by Mohammed Abdul Quddus Khan

Pesticides in Aquatic Environments

Edited by

Mohammed Abdul Quddus Khan
University of Illinois • Chicago, Illinois

Plenum Press · New York and London

Library of Congress Cataloging in Publication Data
Main entry under title:

Pesticides in aquatic environments.

(Environmental science research; 10)
"Proceedings of a symposium for the International Congress of Entomology held in Washington, D.C., August 22, 1976."
Includes bibliographical references and index.
1. Pesticides–Environmental aspects–Congresses. 2. Aquatic ecology–Congresses. I. Khan, Mohammed Abdul Quddus, 1935- II. International Congress of Entomology, 15th, Washington, D.C., 1976. [DNLM: 1. Marine biology–Congresses. 2. Pesticides–Congresses. 3. Water pollution, Chemical–Congresses. 4. Water–Congresses. W1 EN986F v. 10/WA240 P4785 1976]
QH545.P4P482 632'.95042 77-5380
ISBN 0-306-36310-0

Proceedings of a Symposium for the International Congress of Entomology held in Washington, D.C., August 22, 1976

A Division of Plenum Publishing Corporation
227 West 17th Street, New York, N.Y. 10011

Printed in the United States of America

PREFACE

Water covers about two-thirds of the surface of earth, but only 0.627 percent of this water is the sweet surface and subsurface water available for the survival of freshwater organisms including man (1, 2). Some of this fresh or sweet water lies in practically uninhabitable regions (rivers: Mackenzie in Canada; Amazon in Central America; Ob, Yenesey, and Lenta in Siberia, etc.). Also, most of the major rivers (the Mississippi in U.S.A., the Rhine in Europe, the Volga in U.S.S.R., the Ganges in India, etc.), because they flow through agricultural land or urban and industrial areas, have become highly contaminated with chemicals (3). This leaves us with shrinking resources of sweet surface water. In the United States, the dependable supplies of this water are already dwindling in cities like New York and Los Angeles and states like New Mexico and Texas (3).

The current rates of population growth, food and fiber production, industrialization and urbanization will all demand increased use of synthetic chemicals such as pesticides, fertilizers, industrial by-products, etc. This will result in continued contamination of air, land, and water resources. Out of various sources of water pollution this symposium addresses itself to the agricultural chemicals, namely, pesticides. The fate of pesticides in aquatic environments has been a subject of international concern. Scientists and administrators in universities, industry, and in state, federal and international agencies need to sit down together to find solutions concerned with the use, fate, and effects of pesticides. It was with this hope that I approached scientists in various organizations in this and other countries. Unfortunately, the program of the symposium had to be finalized in less than a month. Also, the XV International Congress of Entomology allocated only six hours of time for this symposium. These two factors plus my limited resources made it impractical for me to invite several other scientists whose contribution to the understanding of the fate of pesticides in aquatic environments has been much respected.

It is hoped that the topics covered in this volume will bring to the reader the most up-to-date information on the fate of pesticides in aquatic environments. The book is divided into the following sections:

1. Dynamics of pesticides in aquatic environment: this covers the sources and present state of pesticidal contamination of bodies of water; the physicochemical characteristics that determine the fate of pesticides in water and its absorption by aquatic organisms. Most of the information applies to the North American continent.

2. Dynamics of pesticides in aquatic organisms: this section covers the factors responsible for the pick-up and bioconcentration of pesticidal chemicals. Unfortunately, Dr. B.T. Johnson, who spoke on the absorption of pesticides and the role of trophic migration, decided not to publish his material in this book and this left a vacuum in this section. However, a good deal of information on this topic can be obtained by an article by Kenaga (4). This section also includes the data on elimination of absorbed pesticides by aquatic animals. A paper on the elimination of chlordanes that was presented at the Congress (5) has been included, hoping that this will compliment Dr. F. Matsumura's article.

3. Degradation of pesticides and other foreign chemicals by aquatic organisms: the literature on the metabolism of pesticides by aquatic animals has been reviewed. Again, unfortunately, Dr. D.O. Kaufman decided not to publish his presentation on "Degradation of Pesticides by Aquatic Microorganisms" and due to the shortage of time it was not possible to substitute an article by a leading scientist on this topic. This section thus includes *in vivo* and *in vitro* pathways that pesticides, drugs, carcinogens, and juvenoids undergo in aquatic animals both freshwater and marine.

It is not possible to cover all aspects of the fate of pesticides in one symposium or one book. More similar symposia and publications can lead us to the understanding of the phenomena and processes that are involved in predicting the fate of chemicals in aquatic environments. Several previous symposia and their proceedings on this and related subjects (6-18) can be very useful to the reader.

The success of the symposium and the publication of the proceedings depend mainly on the efforts of the participating scientists. It is for their continued dedication to broaden our knowledge that the students of environmental toxicology in particular and mankind in general are deeply indebted and owe their gratitude.

The participation of Drs. C.A. Edwards (England), F. Korte (W. Germany), and V.H. Morley and J.F. Payne (Canada) is what made the symposium an international effort. I express thanks for their support. I am very grateful to Dr. R.M. Hollingworth for his continued guidance and support to organize this symposium, without which this would not have been possible. Drs. R. Haque (U.S.E.P.A.) and D. Whitacre (Velsicol Chemical Corp.) were very kind in providing links with participants as I was stationed in a new and rather isolated laboratory in Marineland, Florida (C.V. Whitney Marine Research Laboratory of the University of Florida). Although no financial or official support was requested from National Institute of Environmental Health Sciences, the kindness of the institute, especially the encouragement by Drs. R.L. Dixon, L.G. Hart, J.R. Bend, and H.B. Matthews, provided enough stimulation for this undertaking. Dr. R.H. Adamson's (N.I.H.) absence due to his being out of town during the symposium was much felt. I wish to express gratitude to Jennifer Neuman for taking personal interest in typing, correcting, and editing.

As can be judged from the contents of this book, more knowledge and therefore more research is needed to understand the fate of pesticidal chemicals in aquatic environments. The research support for this type of research to individual investigators in universities comes mainly from the National Institute of Environmental Health Sciences, although contractural research is also supported by the Environmental Protection Agency, U.S. Department of Agriculture, and U.S. Department of Interior. The N.I. E.H.S. has the finest review and evaluation procedures (N.I.H.) for processing the grant applications. I personally feel that U.S. Congress and N.I.H. should increase support for these programs sponsored by N.I.E.H.S. and other agencies if the solution of problems related to chemical contamination of aquatic environments is sought.

M.A.Q. Khan

REFERENCES

1. Nace, R.L. 1967. Environ. Sci. Technol. 1: 550.

2. Nace, R.L. 1960. Water Management, Agriculture and Groundwater Supplies. U.S. Geol. Survey Circ. 415, 11 pp.

3. von Hylekama, T.E.A. 1972. In: Environment, Resources, Pollution and Society. (W.W. Murdoch, ed.). Sinauer Assoc. Inc., Stanford, Connecticut. 135-155 pp.

4. Kenaga, E.E. 1975. In: Environmental Dynamics of Pesticides. (R. Haque and V.H. Freed, eds.). Plenum Press. 217-274 pp.

5. Khan, M.A.Q., R. Moore, G. Reddy, R.H. Stanton, and E. Toro. 1976. Paper presented at the XV International Congress of Entomology, Washington, D.C. Aug. 1976.

6. Kearney, C.P. 1973. Significance of Pesticide Metabolites. A symposium organized by the Amer. Chem. Soc., Chicago, Illinois, Aug. 1973.

7. Ecological Toxicology Research Effects of Heavy Metals and Organohalogen Compounds. (McIntyre, A.D., ed.). Plenum Pub. Co. 316 pp. (1975).

8. Environmental Dynamics of Pesticides. (R. Haque and V.H. Freed, eds.). Plenum Pub. Co. 387 pp. (1975).

9. Environmental Pollution by Pesticides. (C.A. Edwards, ed.). Plenum Pub. Co. 542 pp. (1973).

10. Sublethal Effects of hoxic Chemicals on Aquatic Animals. (J.H. Koeman and J.J. T.W.A. Strik, eds.). Elsevier Sc. Pub. Co. 234 pp. (1975).

11. Organochlorine Insecticides: Persistent Organic Pollutants. (F. Moriarty, ed.). Acad. Press. 302 pp. (1975).

12. Survival in Toxic Environments. (M.A.Q. Khan and J.P. Bederka, Jr., eds.). Acad. Press. 550 pp.

13. Environmental Toxicology of Insecticides (F. Matsumura, G.M. Boush, and T. Misato, eds.). Acad. Press. 637 pp. (1973).

14. Chemical and Toxicological Aspects of Environmental Quality: Ecological & Toxicological Aspects of Organochlorines. International Symposium organized by the Institut für ökologische Chemie of the Gesellschaft für Strahlen-und Umweltforschung mbH, München, W. Germany. Sep. 1975.

15. C.A. Edwards. 1974. Persistent Pesticides in the Environment. CRC Press. 170 pp (1974).

16. G.T. Brooks. Chlorinated Insecticides. Vol. 2. Biological and Environmental Aspects. CRC Press. 197 pp. (1974).

17. Bound and Conjugated Pesticide Residues. (D. O. Kaufman, G.G. Still, G.D. Paulson, and S.K. Bandal, eds.). Amer. Chem. Soc. 396 pp. (1976).

18. Air Pollution from Pesticides and Agricultural Processes. (R.E. Lee, ed.). CRC Press. 300 pp. (1976).

CONTENTS

SECTION III

DEGRADATION OF PESTICIDES BY AQUATIC ORGANISMS

L. G. Hart, Chairman

CONTRIBUTORS

Dr. J.R. Bend, National Institute of Environmental Health Sciences, Research Triangle Park, North Carolina

Dr. G.E. Blau, Dow Chemical USA, Midland, Michigan

Dr. G.M. Booth, Department of Zoology, Brigham Young University, Provost, Utah

Dr. T.W. Duke, Director, U.S. Environmental Protection Agency, Gulf Breeze, Florida

Dr. C.A. Edwards, Rothamsted Experiment Station, Herps England

D. Ferrell, Thompson-Hayward Chemical Co., Provost, Utah

Dr. J.R. Fouts, Scientific Director, National Institute of Environmental Health Sciences, Research Triangle Park, North Carolina

Dr. V.H. Freed, Head, Department of Agricultural Chemistry, Oregon State University, Corvallis, Oregon

Dr. R. Haque, United States Environmental Protection Agency, Washington, D.C.

Dr. L.G. Hart, Acting Chief, Pharmacology Branch, National Institute of Environmental Health Sciences, Research Triangle Park, North Carolina

Dr. M.O. James, National Institute of Environmental Health Sciences, Research Triangle Park, North Carolina

Dr. B.T. Johnson, Fish-Pesticide Research Laboratory, U.S. Department of Interior; Fish and Wildlife Service, Rt. 1, Columbia, Missouri

Dr. D.O. Kaufman, Pesticide Degradation Laboratory, U.S.D.A., Beltsville, Maryland

Dr. P.C. Kearney, Chief, Pesticide Degradation Laboratory, U.S.D.A., Beltsville, Maryland

Dr. M.A.Q. Khan, Department of Biological Sciences, University of Illinois at Chicago Circle, Chicago, Illinois

Dr. F. Korte, Institut für Ökologische Chemie, Gesellschaft für Strahlen- und Umweltforschung mbH, München, W. Germany

Dr. F. Matsumura, Department of Entomology, University of Wisconsin, Madison, Wisconsin

Dr. R.L. Metcalf, Department of Entomology, University of Illinois, Urbana-Champaign, Illinois

Dr. V.H. Morley, Research Coordinator, Agriculture Canada, Ottawa, Canada

Dr. W. Brock Neeley, Environmental Sciences Research, Dow Chemical USA, Midland, Michigan

Dr. J.F. Payne, Fisheries and Marine, Environment Canada, St. John's, Newfoundland, Canada

PESTICIDES IN AQUATIC ENVIRONMENTS
AN OVERVIEW

Thomas W. Duke

INTRODUCTION

I would like to add my welcoming remarks to those of Dr. Khan. A glance at the program we are now beginning indicates that we will be privileged to hear from leaders in the area of research on pesticides in the aquatic environment. I was asked to deliver the keynote speech for the symposium and, as I understand it, my purpose is to present a general review of the symposium topic and to introduce the subjects that the speakers will address in more detail as the symposium progresses. Also, I hope to challenge those present to consider certain research issues relating to the program. I will attempt to accomplish these objectives in as few minutes as possible in order to give the speakers the maximum amount of time for their subjects.

TRANSPORT AND FATE

When considering the transport and fate of pesticides in the aquatic environment, one must consider the various kinds of aquatic environments, as well as the many varieties of pesticides. I believe it is appropriate at this time to discuss the terms _aquatic environments_ and _pesticides_. Aquatic environments include such distinct parts of the environment as rivers, streams, lakes, estuaries, coastal and deep ocean waters. Similarities and differences exist among these various parts of the aquatic environment. In each, for example, animals and plants are exposed to and at the mercy of the water media which surrounds them throughout their lives. Conversely, there are obvious differences in the fate and transport of pesticides in freshwater lake systems that are essentially closed and in estuarine systems that are by definition open systems that receive freshwater from river runoff and saltwater from open ocean. There are even differences within similar parts of the environment, such as the saltwater aquatic environment. We speak of a complex foodweb in estuarine and coastal environments and a more simplified foodchain in the open ocean. Pesticides may be considered as ordinary chemical compounds with special characteristics that affect their fate and effect in aquatic environments. Once in the environment, pesticides are exposed to reagents such as oxygen, water, and sunlight and enter the biogeochemical cycles that are continually in operation in aquatic environments. Pesticides exist in a variety of chemical and biological forms including organochlorine, organophosphates, carbamates, juvenile hormones, and viruses. Each of these different chemicals and natural products may interact differently in the various aquatic environments. This is especially true in freshwater and saltwater systems because of physical and chemical differences between waters. How then can we conduct research on the transport and fate of so many different kinds of pesticides and multitudes of formulations in many different aquatic environments? Obviously it is impossible to study every aspect of the problem but we can:

(1) search for common factors or methods of grouping like variables;

(2) concentrate on specific factors that are vital and vulnerable processes, such as sensitive processes within an ecosystem, or simply utilize sensitive indicator species.

Both of the above approaches are necessary and actually are being utilized by investigators; this will be reflected in the papers to be presented later.

Transport and fate of pesticides in the aquatic environment can be affected by at least three main factors: concentration, dilution, and degradation as indicated in Figure 1. (The chart is based on a diagram by Ketchum 1967). Concentration is a form of

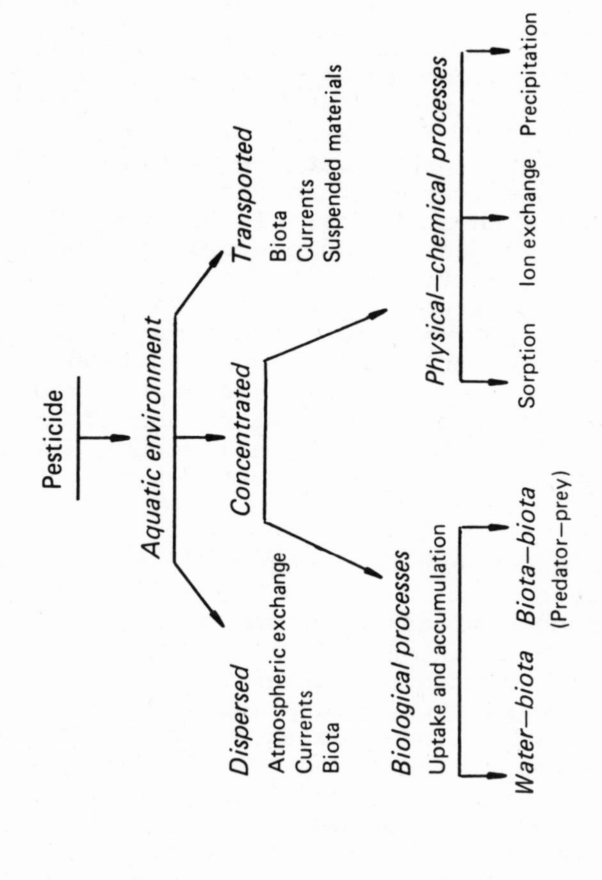

Figure 1. Movement of Pesticides in the Aquatic Environment (Based on Ketchum 1967).

accumulation in a specific component in the environment. For example, bioconcentration occurs when plants or animals accumulate a pesticide to a greater concentration that occurs in the water in which they live. The chemical may be concentrated directly from the water by the organism or it may be transferred from organism to organism through the foodweb to higher trophic levels. The latter phenomenon is especially important when considering the relatively high residues of some pesticides obtained through predator-prey relationships such as amounts of DDT in fish-eating birds.

Pesticides can also be concentrated in the environment through biogeochemical processes. Chemicals are often scavenged from the water through sorption onto suspended material that later is deposited on the bottom to become part of the botton substrate. Consequently, bottom sediments often become reservoirs of pesticides in the environment. In marine waters, especially, pesticides may be precipitated from water through chemical reactions. This can result in the accumulation of pesticides at the mouth of an estuary where water from a freshwater stream meets saltwater from the sea. It should be noted at this point that pesticides are not necessarily homogeneously concentrated throughout an aquatic system, and it is necessary to evaluate areas of "critical accumulation," such as critical organisms or other components of the ecosystem.

Dilution may occur as a result of dispersion of a pesticide throughout a body of water from a point source by biogeochemical reactions. Pesticides accumulated by pelagic organisms can be dispersed quickly and for great distances by the animals as they swim from one area to another. Surprising amounts of pesticides can be transported from an estuary to open ocean through migration of organisms that have concentrated the chemical from their in-shore environment. Pesticides accumulated by suspended materials may also be transported to various areas through erosion and the transport from water to the atmosphere through atmospheric exchange. These kinds of dilution and transport, of course, depend upon local climatic conditions.

Pesticides in the aquatic environment may undergo transformations or degradations due to photochemical reaction, chemical reaction, and biological reactions, or a combination of these phenomena. Whatever the mechanisms of degradation, the parent compound is changed--sometimes to one more compatible with the environment and sometimes to one more harmful to the environment. The field of microbiological degradation of pesticides in the aquatic environment has received some emphasis during the past few years, and its importance, no doubt, merits additional attention. If one is to predict the impact of pesticides on the aquatic environment, one must understand that information should be available on the transformation the pesticide undergoes in reaching its final form as a residue in components of the environment.

The distribution and levels of concentrations of pesticide residues in the aquatic environment is revealed in several existing monitoring programs within the United States and contiguous waters. Waterfowl, fish, and shellfish prove to be excellent species for indicating the presence of these chemicals in the environment. Fish, especially pelagic fish, travel great distances, thus indicating the occurrence of pesticides somewhere within their range. Conversely, shellfish are sedentary and can be used to more specifically locate the source of the residue.

It is necessary to be familiar with the biology of indicator organisms to correctly interpret the meaning of residue levels. For example, organochlorine pesticides concentrate in the gonadel tissue of shellfish and is lost or greatly decreased when spawning occurs. Thus a monitoring sample taken during spawning could erroneously indicate a lack of the residue in the environment--another reason for requiring that frequent samples be taken and analyzed for trends of concentration versus time, as opposed to extrapolating from single monitoring samples.

EFFECTS OF PESTICIDES ON AQUATIC ENVIRONMENTS

The presence of some pesticides can be detected in biological and physical components of aquatic environments at the parts per trillion level, but the effect of such concentrations of pesticides on the organisms and systems in which they occur are not clear in many instances. Knowledge of such effects are especially important in aquatic environments that often interface with man's activities on land and, therefore, are especially suspectible to exposures to acute doses of degradable pesticides, as well as chronic doses of persistent ones. In order to decrease the chances for adverse effects of pesticides on nontarget aquatic organisms, the following guidelines have been suggested in the U.S. Environmental Protection Agency Research Series, March 1973, Water Quality Criteria 1972:

(1) A compound, which is the most specific for the intended purpose, should be preferred over a compound that has broad spectrum effects;

(2) A compound of low persistence should be used in preference to a compound of greater persistence;

(3) A compound of lower toxicity to nontarget organisms should be used in preference to one of high toxicity;

(4) Water samples to be analyzed should include all suspended particulate and solid material: Residues associated with these should therefore be considered as present in the water;

(5) When a derivity such as p,p'-DDE, or 1-Napthol is measured with or instead of the parent compound, deduct toxicity of the derivative should be considered separately: If the toxicity of a derivative, such as an iononic species in a pesticide, is considered equivalent to that of the original parent compound, concentration should be expressed as the equivalence of the parent compounds. These general guidelines are helpful to those applying pesticides to insure the desired action impinges the target organisms and not nontarget, aquatic organisms.

Much of the information now available on the effect of pesticides on aquatic organisms and their environments is in terms of acute and mortality of individual species. These results were often obtained in routine bioassays, where the test organisms were subjected to the pollutant and periodically compared with controlled organisms. During the past few years, the need for information on the chronic or partial chronic exposures on sublethal effects of pesticides on aquatic organisms and ecosystem has become evident. Data are now available on chronic studies involving the exposure of aquatic organisms to pesticides over an entire lifecycle; these studies are often referred to as "egg-to-egg" studies. Sublethal data are available on the effect of concentrations of pesticides less than those lethal to organisms and utilize such criteria for effects as growth, function of enzymes systems, and behavioral populations of organisms. Results of these kinds of effects studies must be coordinated with results obtained at the system or the community or ecosystem level.

Microcosms or experimental environments in the laboratory and in field systems of relatively restricted size have proven useful in effects studies. These test systems are often miniature representatives of specific ecosystems that contain components and processes necessary to investigate particular problems or questions. In addition to being useful in effects studies, they also provide data on potential transport, fate, and degradation of pesticides in the environment. In general, these kinds of studies provide a basis for statements about fate and effects of pesticides in larger systems.

CHALLENGES

I should like to challenge the speakers and audience to interact on the following issues during our symposium:

(1) The state-of-the-art for extrapolating from laboratory studies to the natural environment, including the strong points and limitations of using microcosms and other systems to develop predictive data on fate and effects of pesticides in the aquatic environment;

(2) The possibility of generalizing or placing pollutants in general categories so that "shortcut" predictions can be made when dealing with many different kinds of pesticides, formulations, and aquatic systems;

(3) The need for researching unstressed aquatic systems;

(4) The importance of elucidating the more vital and vunerable processes in aquatic system in order to predict the fate of pollutants and aquatic environments.

REFERENCES

Ketchum, B.H. 1967. Man's Resources in the Marine Environment. In T.A. Olsen and F.J. Burgess, Editors, Pollution and Marine Ecology, Inter-Science Publishers, New York, 1-11 p.

U.S. Environmental Protection Agency Research Series. March 1973. Water Quality Criteria 1972. Section IV, Marine Aquatic Life and Wildlife. U.S. Government Printing Office, Washington, D.C. 269 p.

Section I

PESTICIDES IN AQUATIC ENVIRONMENTS

P. C. Kearney, *Chairman*

NATURE AND ORIGINS OF POLLUTION OF AQUATIC SYSTEMS BY PESTICIDES

Clive A. Edwards

ABSTRACT

Pesticides reach aquatic systems by direct application, spray drift, aerial spraying, washing from the atmosphere by precipitation, erosion and run-off from agricultural land, by discharge of effluent from factories, and in sewage. The relative importance of these sources are discussed and evaluated; it is concluded that run-off from agricultural land is the main source of gradual pollution, with direct application to water and discharge of effluent into aquatic systems causing more serious, but localised contamination. The pesticides that cause most pollution are the organochlorine insecticides and certain herbicides. In water, pesticides become bound to organic matter in mud and sediment quite rapidly, only small amounts remaining in solution. There is a continuous interchange between sediments and water, influenced by water movement, turbulence and temperature. Pesticides are also taken up into the biota but the overall amounts stored in this way are small relative to the overall amounts in aquatic systems. Pesticide residues are largest in rivers, less in estuaries and least in the oceans.

INTRODUCTION

Pollution of aquatic systems is widespread, sometimes being so bad that no fish can live in the polluted

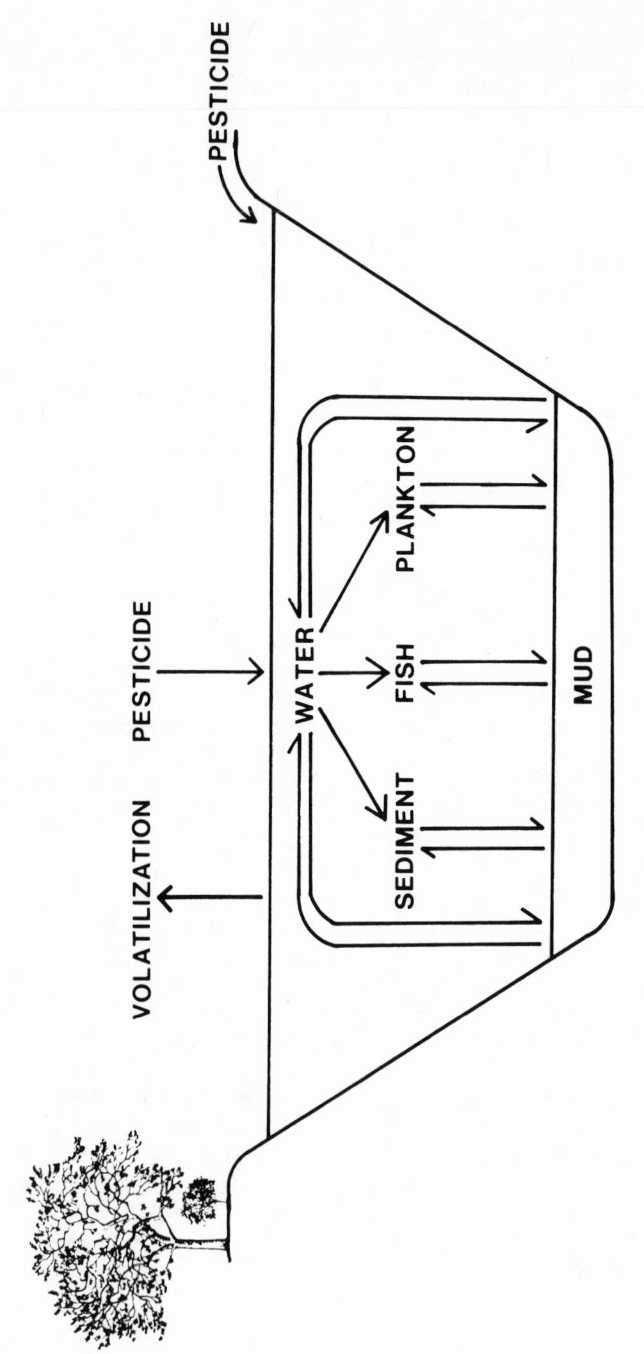

Fig. 1. Movement of pesticides through aquatic systems.

waters; pesticides are one of a wide variety of organic pollutants contributing to this situation. It is important to describe the nature, origins and extent of such pesticide pollution at the beginning of this symposium in order to set the scene for later speakers who will consider the dynamics of pesticides in aquatic environments, aquatic organisms and the way in which these pesticides are degraded.

Pesticides may reach the waters of streams, rivers, lakes and oceans in many different ways, and the types of pesticides involved differ greatly. It is probably true that at some time or other almost every pesticide ever used must have polluted some body of water. It would be outside the scope of this paper to consider all these in detail, so most of the discussion will be confined to pollution by the persistent organochlorine insecticides, and a number of relatively stable organophosphate and carbamate insecticides and herbicides. I hope this will be the pattern followed by the later speakers and that our discussion will help resolve some of the main problems involved in pollution of aquatic environments by pesticides.

I shall first consider the nature of pesticide pollution of aquatic environments, then their origins, followed by a consideration of partitioning amongst the various components of the hydrosphere and, finally, the degree of contamination of various aquatic systems.

NATURE OF PESTICIDE POLLUTION OF AQUATIC SYSTEMS

Three quarters of the surface of the globe is covered with water. Any chemical that contaminates soil or the atmosphere has the potential for transfer to the hydrosphere and pollution of this medium. Pesticides are used in vast and ever-increasing amounts; many of them are relatively persistent, and leave considerable residues in the soil or the atmosphere, and from these two media they are readily transferred to aquatic systems, carried in precipitation or in run-off and drainage from land to water. Once they reach large bodies of water, they are readily transported by diffusion, in water currents, or in the bodies of aquatic organisms, throughout any continuous aquatic system. Once in aquatic systems, pesticides must be either degraded to simpler compounds, remain there, or move back into the atmosphere by volatilisation (Fig. 1). The ultimate fate

Table 1. Characteristics of Pesticides that Cause Hazards to Aquatic Systems

CHARACTERISTIC	INSECTICIDES			
	ORGANOCHLORINES	ORGANOPHOSPHATES	CARBAMATES	HERBICIDES
1. HIGH TOXICITY TO AQUATIC FAUNA	+ + +	+ +	+ +	+
2. SOLUBILITY	-	+	+	+ + +
3. POTENTIAL FOR UPTAKE & BIOCONCENTRATION	+ + +	+	+	+ +
4. PERSISTENCE IN AQUATIC SYSTEMS	+ + +	+	+	+ +
5. HEAVY USAGE ON OR CLOSE TO WATER	+ + +	+ + +	+ +	+ + +

of most persistent pesticides is still not fully understood.

The effects of pesticide pollution of aquatic systems depend not only with the characteristics of the pesticide and its concentration, but also with the nature and biology of the aquatic system. I shall consider these aspects in turn.

Characteristics of Pesticides that Cause Pollution

A great variety of chemicals of extremely different structures and properties have been used as pesticides. They range from simple inorganic compounds such as lead arsenate and copper sulphate to complex mixtures of organic molecules. Some are broad spectrum biocides, others are very toxic to particular groups of animals or plants and yet others are relatively innocuous to nearly all organisms. The solubility of pesticides in water ranges from 0.001 ppm for DDT, to as much as 70% solubility for the herbicide diquat. Their potential for bioconcentration ranges from zero to very high. The chemical stability of pesticides in water can range from a few hours to many years. I shall try to evaluate the relative importance of these characteristics in turn; these are summarized in Table 1.

<u>The stability and persistence of a pesticide in aquatic systems</u> is a function of its chemical structure. Pesticides range in stability from those that are very unstable and break down in a few hours to extremely stable compounds able to persist as residues for many years (Table 2). Clearly, those pesticides that are very persistent, present a potential hazard to the productivity of aquatic systems, because the aquatic biota will be exposed to residues for considerable periods even after only a single contamination, and if such contamination is repeated or occurs continuously, there is potential for accumulation of the more persistent pesticide in some components of an aquatic system. Our knowledge of the relative persistence of different pesticides in aquatic systems is still scanty. Some data on residues in fish are given in Table 3. Data on residues in water tend to vary markedly with season, the degree of turbulence of the water, and amount of suspended particulate matter, so fish are often considered to be a much better indication of water pollution by pesticides than analyses of water samples, because the residues in their tissues

Table 2. Persistence of Pesticides in River Water

(Adapted from Eichelberger and Lichtenberg, 1971)

Pesticide	Percentage remaining	
	2 weeks	4 weeks
Organochlorines		
BHC	100	100
Dieldrin	100	100
DDT	100	100
Endrin	100	100
Chlordane	100	85
Aldrin	80	40
Endosulfan	30	5
Isobenzan	25	10
Heptachlor	25	0
Heptachlor epoxide	100	100
Organophosphates		
Monocrotophos	100	100
Dimethoate	100	85
Ethion	90	75
Fenthion	50	10
Parathion	50	30
Malathion	25	10
Trithion	25	10
Carbamates		
Aminocarb	60	10
Propoxur	50	30
'Zectran'	15	0
Carbaryl	5	0
Fenuron	60	20
Monuron	40	30

Table 3. The persistence of pesticides in fish

(From data by Macek, 1970)

Pesticide	Persistence in fish
Herbicides	
Diquat	< 3 weeks
Endothal	< 3 weeks
Simazine	< 3 days
2,4-D	< 1 week
Sodium arsenite	> 16 weeks
Dichlobenil	< 2 weeks
Organophosphate Insecticides	
Diazinon	< 1 week
Dursban	< 1 week
Azinphos-methyl	< 1 week
Malathion	< 1 day
Parathion	< 1 week
Organochlorine Insecticides	
Lindane	< 2 days
Methoxychlor	1 week
Heptachlor	1 month
Dieldrin	1 month
Camphechlor	> 6 months
DDD	> 6 months
DDT	> 5 months

are several orders of magnitude higher and much easier to analyse. Fish are sensitive indicators of residues in water in which the amounts of pesticides present are too small to be reliably analysed chemically. Persistence in the total aquatic system is greatest for organochlorine insecticides, intermediate for organophosphate and carbamate insecticides and least for herbicides.

High toxicity to the aquatic fauna or flora is an important criterion for a pesticide to be important as a pollutant of aquatic systems. No matter how persistent a pesticide may be in water, if it causes no serious harm, it cannot be considered to be an important pollutant.

Some pesticides are extremely toxic to fish at very low concentrations and to aquatic invertebrates at even lower concentrations. Toxicity may be confined to a small group of organisms and thus be highly specific, or at the opposite extreme, affect almost all forms of plant and animal life in water. More usually, pesticides differ in toxicity to different aquatic organisms. Toxicity may be acute and kill the organism relatively quickly, or chronic and have gradual effects upon activity, feeding, reproduction and general physiology; the first is normally much more obvious because mortality of large numbers of fish soon attracts attention; the latter is much harder to detect. The relative toxicity of some pesticides commonly polluting aquatic systems are given in Table 4, although these laboratory assessments of toxicity do not always agree with those in field experiments.

High water solubility is a characteristic which can add or detract to the potential of a pesticide as a pollutant of aquatic systems. Pesticides differ greatly in solubility (Table 5). If a pesticide is relatively soluble, it can drain into aquatic systems more readily from agricultural land and permeate rapidly through the whole of the system. However, more soluble pesticides tend to be less persistent in water and more easily diluted. By contrast, insoluble pesticides are not readily leached into aquatic systems and once in a system are rapidly bound to living or dead organic matter or fractions of the bottom sediment. Therefore, they affect the biota only if the chemical is very toxic, accumulates in the tissues of organisms or is bound on to organic matter that provides food for other organisms.

Table 4. Toxicity of Pesticides to Fish

Insecticide	L.C.$_{50}$ (ppm)*	
	Bluegills	Rainbow Trout
ORGONOCHLORINES:		
Endrin	0.006	0.007
Toxaphene	0.004	0.008
Dieldrin	0.008	0.019
DDT	0.016	0.018
Aldrin	0.013	0.036
Chlordane	0.022	0.022
Methoxychlor	0.062	0.020
Lindane	0.062	0.060
BHC	0.79	-
Heptachlor	0.190	0.150
ORGANOPHOSPHATES:		
Azinophosmethyl	0.022	0.014
Phosdrin	0.041	0.034
Fonofos	0.045	0.110
Malathion	0.120	0.100
Diazinon	0.052	0.380
Phosphamidon	-	5.0
Methyl parathion	5.7	2.7
Dimethoate	28.0	20.0

Table 4 - Continued

	Bluegills	Rainbow Trout
CARBAMATES		
Carbaryl	3.4	3.5
'Zectran'	11.2	10.2
Herbicide		
Trifluralin	0.019	0.011
2,4-D	0.9	1.1
Fenoprop	16.6	1.4
Diuron	7.4	4.3
Chlorfenac	19.0	7.5
Diquat	19.0	20.0
Dichlobenil	20.0	22.0
Sodium arsenite	44.0	36.5
Paraquat	-	62.0
Simazine	130.0	85.0
MCPA	100.0	-
Fenoprop	120.0	-
Aminotriazole	100.0	-
Dalapon	115.0	-

*96 hr for organochlorines, 24 hr for organophosphates and carbamates, and 48 hr for herbicides

Table 5. SOLUBILITY OF PESTICIDES IN WATER (ppm)

Pesticide	Solubility
ORGANOCHLORINES	
DDT	0.0012
Aldrin	0.01
Heptachlor	0.056
Methoxychlor	0.10
Dieldrin	0.18
Endrin	0.23
Toxaphene	0.40
Lindane	7.0
ORGANOPHOSPHATES	
Parathion	24
Disulfoton	25
Azinphosmethyl	30
Diazinon	40
Phorate	50
Chlorfenvinphos	145
Malathion	145
Demeton methyl	330
Thionazin	1,140
Dimethoate	2,500
CARBAMATES	
Carbaryl	40
Carbofuran	700
Aldicarb	6,000
HERBICIDES	
Simazine	5.0
Propazine	8.0
Dichlobenil	20
Atrazine	33
Diuron	42
Chlorfenac	200
2,4,5-T	280
Monuron	280
2,4-D	890
Aminotriazole	28%
Diquat	70%
Dalapon	80%

Potential for uptake and bioconcentration into aquatic organisms is an important characteristic in determining how serious the effects of a particular pesticide on an aquatic system can be. Many pesticides are able to accumulate from water into the tissues of animals and plants; this is especially a feature of most of the organochlorine insecticides which often reach concentrations many thousand times greater in the biota than in the water.

One of the basic properties of a pesticide that causes accumulation, is a high lipoid/water partition coefficient, which can be determined readily. Similarly, the degree of bioconcentration can be assessed by collecting organisms from contaminated systems and measuring the residues that they and the medium in which they live contain. Nevertheless, the mechanisms by which such bioconcentration occurs is often uncertain, depending on many factors such as time of exposure, rate of uptake, metabolism within the organism, rate of excretion, potential for storage and physiological state of the organism.

The process of bioconcentration from the medium in which organisms live is sometimes confused with that of biological magnification, which can be defined as the accumulation of a pesticide in an animal in any particular trophic level at a concentration greater than that in its food or the preceding trophic level so that eventually, animals at the top of food chains accumulate the largest residues. This concept is an extremely widely held one, and, since the publication of "Silent Spring" (Carson, 1962), has been commonly accepted as valid.

However, it seems probable that this concept is greatly over-simplified and there are many other possible explanations for the tendency for higher residues to occur in predators than in their prey, food chains being only one factor (Moriarty, 1973; Rosenberg, 1975). The food chain concept is particularly questionable in aquatic systems because, although there is a marked tendency for organisms in the higher trophic levels to accumulate larger amounts of persistent organic pesticides, there is good evidence that these organisms can obtain these residues as readily from the water in which they live as from their food. For instance, when three organisms in a food chain (an alga, *Scenedesmus obliquus*, a crustacean, *Daphnia magna* and a fish, *Poecilia reticulata*) were exposed separately to dieldrin in a continuous flow system the concentration factors (concentration in the dry weight of organism divided by concentration in water) were about 1,300 for the alga, 14,000 for the crustacean and 49,000 for the fish. By comparison, contaminated Daphnia fed to the fish could produce only one tenth of these residues (Reinert, 1972). Similar results were obtained by Chadwick and Brocksen (1969).

Heavy usage of a pesticide, especially in, or close to aquatic systems is an important factor in the degree of pollution of any system by a particular pesticide. In particular, application of pesticides as sprays, lead to drift and atmospheric contamination and can cause considerable fallout on to systems with a large surface area.

There is strong evidence, from many sources, that there is much more contamination of aquatic systems close to areas of heavy pesticide use. For instance the frequent contamination of the Mississippi River by endrin, although partially due to effluents, is largely caused by the use of this insecticide on cotton. When a 400 square mile agricultural watershed in Alabama with a cotton acreage of 13,000 to 16,000 acres was studied intensively for pesticide residues, there was a strong correlation between amounts and types of residues in the water and annual and seasonal variations in use.

Characteristics of Aquatic Systems that Affect Pollution by Pesticides

The size of aquatic systems can range from tiny ponds to whole oceans, so this is obviously a major factor in determining their susceptibility to pollution by pesticides. A partly used can of pesticide disposed in a farm pond can have a catastrophic effect on the system; in an ocean it would be rapidly diluted and have no appreciable effect. Conversely, atmospheric fall-out over a large area may cause serious pollution if most of the pesticide becomes concentrated into a small part of the biota, but might have little effect on a small pond. When a large river becomes polluted by pesticides the results can be very serious with extremely large numbers of fish killed, but contamination of a small stream is much less important. The possible extent of such incidents were seen when the Mississippi was polluted by endrin (U.S.D.A., 1966) and the Rhine by endosulfan (Greve, 1972). Similar pollution of a small stream would cause little interest.

The form of an aquatic system can also greatly affect its potential to become polluted by pesticides. If it is a stagnant system containing little oxygen and much partly decayed organic matter, the pesticide becomes bound on to the organic matter, localized in distribution and may be degraded anaerobically more rapidly than it would be in a well-oxygenated system. In a flowing system such as a river the pesticide becomes more widely distributed and diluted.

The depth is also of considerable importance; in a deep system most of the pesticide (unless it is very soluble) becomes bound on to sediment at the bottom and has little influence on the surface biota; in a shallow system, most of the biota are continually exposed to the pesticide. For instance, Terriere et al. (1966) reported that camphechlor persisted for about one year in a shallow lake, but up to five years in a much deeper lake with less biological activity.

The location of an aquatic system is extremely important in influencing its potential for pesticide pollution. On a global scale, much greater quantities of pesticides are used in N. America and Europe than in other regions so aquatic pollution by these chemicals is more likely in these areas.

Climate also exerts an influence; where there is considerable precipitation much larger quantities of pesticide residues may be swept from the atmosphere and deposited on the surface of bodies of water. High temperatures favour volatilization of pesticides from the soil surface or from the surface of water.

On a more localized scale, if an aquatic system is close to agricultural land to which large amounts of pesticides are applied, the potential for contamination is much greater than in an area several miles away. Similarly, there is often aquatic pollution by pesticides in urban areas where large amounts are used for domestic purposes.

ORIGINS OF PESTICIDE POLLUTION OF AQUATIC SYSTEMS

Aquatic systems become polluted by pesticides either by deliberate application or accidentally; the latter being more usual. Accidental contamination occurs either insidiously by spray drift, in precipitation, in drainage and run-off, or precipitously, when large amounts of industrial effluent reaches water or when waste pesticides or containers are dumped into water. Gradual contamination of the former type carries greater potential environmental hazard, because it is not dramatic, and may be completely overlooked. Dramatic fish kills become immediately obvious, there is considerable public outcry and the causes are usually eliminated or controlled. Table 6 summarizes the relative importance of some of these sources of pollution on different aquatic systems.

Table 6. AQUATIC SYSTEMS THAT MAY BE POLLUTED BY PESTICIDES

SYSTEM	MAIN SOURCES OF POLLUTION			
	SPRAYING	RUN-OFF	ATMOSPHERE	EFFLUENT
PONDS	+ +	+ +	+	-
RIVERS	+	+ +	+	+ + +
ESTUARIES	+	+	+	+ +
OCEANS	-	+	+ + +	+

Application of Pesticides to Aquatic Systems

Pesticides are applied to aquatic systems for a variety of reasons and in many different ways, such as for:

Fish eradication in lakes, in order to restock with more desirable fish which is a common, if poorly-known practice. The pesticide used for this purpose for many years was rotenone and this caused little hazard. More recently sprays of camphechlor ('Toxaphene') have been used for this purpose and it is ironical that this is an organochlorine insecticide that has been accused of serious side-effects when reaching aquatic systems accidentally. It is the most toxic organochlorine to fish except for endrin (Table 4) and is fairly persistent in water, with up to twelve months being necessary before restocking with other fish is safe.

Another chemical used for this purpose is 'Antimycin', an antibiotic which is to some extent selective. It is applied on the surface of a carrier which allows controlled release of the toxicant (Muirhead-Thomson, 1971).

Control of lampreys, which are serious predators of fish, using a chemical (tri-fluoro methyl nitrophenol) in the Great Lakes has met with considerable success because of the selective action of the chemical: nevertheless, the margin of toxicity between lampreys and desirable fish is small and there is some potential for side-effects to occur.

Control of aquatic snails that carry various diseases such as *Bilharzia* of man, or liver fluke in domestic animals requires application of various types of chemicals to aquatic systems. More usually, treatment is applied mainly to the perimeter of aquatic systems but sometimes to whole bodies of water. Fortunately, molluscicides such as *N*-trityl morpholine ('Frescon') are relatively specific and of little harm to other organisms, although some such as copper sulphate, niclosamine, and sodium pentachlorophenate (NaPCP) can kill fish and have sometimes caused serious aquatic problems.

Control of insect pests such as blackfly (*Simulium* spp.) mosquitoes (*Anopheles*, *Culex* and *Aedes*) and non-biting midges (Chironomidae) and gnats (Chaeoboridae) all of which have aquatic larvae, has been most commonly by DDT, although other organochlorine and organophosphorus insecticides have also been used. In rivers, such treatments lead to considerable contamination for miles down-stream, and in lakes many groups of animals are affected by the pesticide residues. The most noteworthy case was when TDE was used to control gnats in Clear Lake, California resulting in considerable accumulation in aquatic organisms so that certain fish-eating birds were killed (Hunt and Bischoff, 1960).

Other chemicals used to control aquatic insects include BHC, EPN, fenthion, DDVP, malathion, trichlorphon, chlorpyrifos ('Dursban') and many others. All of these are relatively non-specific and moderately toxic to fish, so severe disruptions of aquatic systems are possible.

Control of weeds, which choke many static aquatic systems, using aerial sprays of herbicides has become increasingly common in recent years. Because these weeds belong to a wide range of plant families and species, many different herbicides have been used. Fortunately, herbicides are only moderately or slightly toxic to fish (Table 4) but there is considerable variation in susceptibility between different species of fish.

Herbicides commonly used in aquatic weed control, include diquat, paraquat and 2,4-D, 2,4,5-T, endothal, fenac and aminotriazole. Although, these are only moderately toxic to fish and not very persistent, fish kills have been reported after their use.

Spray Contamination by Drift

It is rare for a large proportion of a pesticide sprayed on to a crop to reach its target; commonly, as much as 50% may be lost, either as drift of larger particles or by volatilization of smaller particles during spraying. This drift may be carried considerable distances by wind or air currents but usually the larger particles containing most of the pesticide residues fall out close to the site of application and can be a source of contamination of aquatic systems close to spraying operations. Some indication that aquatic systems are being polluted in this way can be obtained when there are distinct seasonal fluctuations in pesticide residues in water or bottom sediment. Greater contamination is possible by drift from aerial spraying operations. Hindin et al. (1966) reported that less than 35 percent of DDT and other pesticides sprayed from aircraft reached their target; presumably most of the rest was carried into the atmosphere. Aerial spraying of forest areas with persistent pesticides to control such pests as spruce bud worm and gypsy moth has caused considerable contamination of forest streams in the past, but does not seem to have led to any serious long-term effects. For instance young salmon were killed in the Miramichi River, Canada after DDT was applied to forest for spruce bud worm control. Other instances have been in Maine, Columbia and Montana (Faust, 1964).

Contamination from Treated Land-Drainage

Agricultural land usually drains into ground water or some aquatic system, and movememt of pesticides would seem a likely source of contamination of such systems. However, many pesticides are relatively insoluble and soil is not only an extremely efficient filter, but also strongly binds many organic chemicals as they percolate through it so that only the most soluble and unreactive pesticides reach drainage waters in appreciable amounts. For instance, we found (Thompson et al., 1970) that less than 2 percent of dieldrin was leached through soil after a years rainfall and Terriere et al. (1966) stated that less than 0.1 percent of DDT applied to orchards reached ground water. However, for soluble pesticides and where the amount of pesticide is large, chemicals such as 2,4-D can contaminate ground-water and wells (Walker, 1961). The soil type is an important factor, much more pesticide leaching through sandy soils low in organic matter, than through clays, peat and muck soils (Edwards, 1966).

Contamination by Run-Off from Treated Land

Another source of pollution of aquatic systems is by direct surface run-off. This can occur directly, with relatively soluble pesticides, or, for the more insoluble pesticides, bound on to particulate matter suspended in the run-off waters. There are many variables other than solubility that influence the amounts of pesticides that may reach aquatic systems by this route. These include: the amount of suspended solid organic matter in the run-off water, the sorption characteristics of the solids, the distance between treatment areas and water, the slope and vegetative cover of the land, the time that elapses between application and precipitation and the amount of rainfall. The amount of a pesticide such as DDT that reaches water may be as small as 0.01 percent of the amount applied on relatively flat agricultural land (Hindin et al. 1966). Even on land sloping moderately towards water, amounts may not be large as was shown for dieldrin (Edwards et al., 1970). However, much larger quantities may move in this way in watersheds with a large amount of rainfall. The optimum condition for run-off is when there is heavy rainfall soon after a pesticide is applied. This is because most pesticides become progressively absorbed on to soil fractions. For instance, in one experiment 3.4 kg/ha of atrazine was applied to land; after 6.3 cm of rain fell within an hour of treatment 17 percent was lost compared with only 7 percent when rain occurred 4 days after treatment (White et al., 1967). One characteristic of pesticides that enter water bound to particulate matter is that they reach the bottom sediment rapidly, and pollution in this way may cause fewer side-effects than when the pesticides enter aquatic systems in solution.

Contamination from Treated Land-Erosion

Pesticides become tightly bound to various soil fractions so that pesticide residues that originate from fall-out of sprays on to soil are mostly concentrated close to the soil surface. In dry weather, such deposits may be blown away as dust, and deposited at considerable distances from their point of origin. Clearly, the destination of some of these residues will be open bodies of water either as dust fall out or when the particles are carried down in precipitation. However, it seems unlikely that large residues will reach aquatic systems in this way.

Contamination from the Atmosphere in Precipitation

There is good evidence that the atmosphere contains large amounts of various organic and inorganic chemicals including, particularly, persistent pesticides. These are in the vapour phase, absorbed on to dust or in solution in atmospheric water. They reach the atmosphere in many ways including spray drift, volatization from soil and water and dust erosion. During precipitation, such residues in the lower atmospheric strata are swept out and fall on to land or water. Although, this is a continuous source of contamination, the amounts involved are usually very small per unit area, particularly at large distances from treated land, and unless there is a great degree of bioconcentration into aquatic organisms they are unlikely to affect aquatic systems seriously.

Pollution by Discharge into Water

Many aquatic systems are heavily contaminated by a wide range of pollutants, arising from discharge of industrial effluents or sewage into water. The treatments applied to industrial waste range from virtually none to complex detoxification processes (Nicholson, 1969).

Pesticides can reach water in such effluents, either as waste from pesticide factories, or as discharge from factories such as woollen mills that use pesticides in their processing. There are many reports in the literature of contamination of aquatic systems in this way, such as that in Lake Michigan in U.S.A. or in rivers receiving industrial effluents in Yorkshire and Lancashire in England (Lowden et al., 1969).

SITE AND EFFECT OF PESTICIDE

It is important to emphasize that water is only one component of the hydrosphere that may be affected by pesticides and usually contains much smaller amounts of relatively insoluble chemicals than surface slicks, suspended particulate matter, bottom sediment and the tissues of living organisms. When aquatic systems are contaminated by pesticides in solution, there is a strong tendency for the residues to move to other components of the system. This is illustrated by laboratory experiments in which water containing DDT was placed above soil covered with filter paper; after six hours 56 percent of the DDT was in the soil and after 24 hours there was a further 22 percent in the soil (Weidhaas et al., 1960). Thus partitioning of a polluting pesticide is reversible; when there is turbulence or other disturbance the amount in the water may increase again.

It is convenient to consider the main sites and effects of polluting pesticides i.e. water, sediment and the biota separately.

Water

Except in unusual circumstances associated with effluent or spillage, amounts of pesticides in water are usually very low, at most parts per thousand million (U.S. billion, pp 10^9). It should also be noted here that nearly all water analyses of residues are made on unfiltered water, so that residues existing on fine particulate matter are also assayed. In fact, some of the organochlorines are so readily absorbed on to surfaces of container walls that they are almost impossible to measure analytically with accuracy. Evidence that residues reported, are not usually those really in solution was given by a study of pesticide residues on Lake Erie. On a moderately windy day, the concentrations of pesticide ranged from 50 pp 10^9 close to the shore to about

3 pp 10^9 about 10 miles from shore where there was virtually no turbidity (Hartung, 1972).

However, there is a tendency for some of the more stable and more soluble herbicides, that reach water, to remain in solution for long periods. Fortunately, these tend to be less liable to bioconcentration and less toxic to fish.

Mud and Sediment

Since most pesticides contaminating water are absorbed on to particulate matter, and there is continual tendency for such material to sink to the bottom of aquatic systems, it follows that the larger pesticide residues probably occur in the bottom sediments. That does not mean that the greatest concentrations occur in the sediment but rather the greatest overall amounts are deposited there.

The bottom sediment is an extremely variable material containing clay, silicates, organic matter in different stages of decay and other hydrocarbons including lipoid-containing materials of various sorts. These may cause the lipophilic persistent pesticides to partition from water as well as causing absorption of these chemicals on to the organic matter.

Once in the sediment, a pesticide may remain undisturbed for long periods in relatively calm water, depending on the type of pesticide. The anaerobic conditions usually occurring in sediments tend to favour degradation of many organochlorine insecticides, particularly DDT and BHC (Hill and McCarty, 1967).

Plankton and Floating Material

Nearly all fresh-water or marine aquatic systems contain large numbers of microorganisms and small animals that float near the water surface. These are often associated with floating organic particulate matter. The nature of such plankton, which is a major food source for fish, differs greatly in different systems. Its abundance and variety depends on the mineral and organic content of the water, oxygenation and amount of organic and inorganic pollution.

Pesticides may kill plankton, affect growth of individuals and populations, and even photosynthesis.

Plankton animals and plants are extremely susceptible to organochlorine insecticides and to phenylurea herbicides such as monuron and diuron, and, to a lesser extent, organophosphate insecticides such as chlorfenvinphos. When plankton is killed by a pesticide sprayed on to the surface of water, they tend to fall rapidly to the bottom, so that there is spectacular clearing of the water. The uptake and fall-out tends to remove the pesticide from the surface waters and take it down to the bottom.

Plankton can also readily take pesticides up into their tissues although, usually to a lesser degree than fish. Much of the uptake seems to be passive depending mainly on lipid/water partition coefficients, because dead plankton will often accumulate as much as live organisms of the same species.

Fish

The gills of fish are extremely efficient in removing persistent pesticides from water. For instance, Atlantic salmon (Salmo solar) exposed to one ppm DDT in solution (greater than its water solubility) concentrated 1.5 ppm DDT in their livers and spleens in five minutes and 31 ppm in one hour (Premdas and Anderson, 1963). There are many similar reports, and it seems clear that the main uptake of pesticides by fish from water is direct but passive, and the entry into fish with their food is much less important.

Such uptake of pesticides into fish from water continues until an equilibrium level is reached; this level depends on the concentration in the surrounding water and the time taken to reach equilibrium depends upon the nature of the pesticide, species of fish and other environmental factors; it may take from a few hours to several weeks for equilibrium to be attained (Kenaga, 1972). When fish are transferred to clean water, the pesticide residues are gradually excreted, usually in two phases; an initial rapid phase followed by a much slower gradual loss.

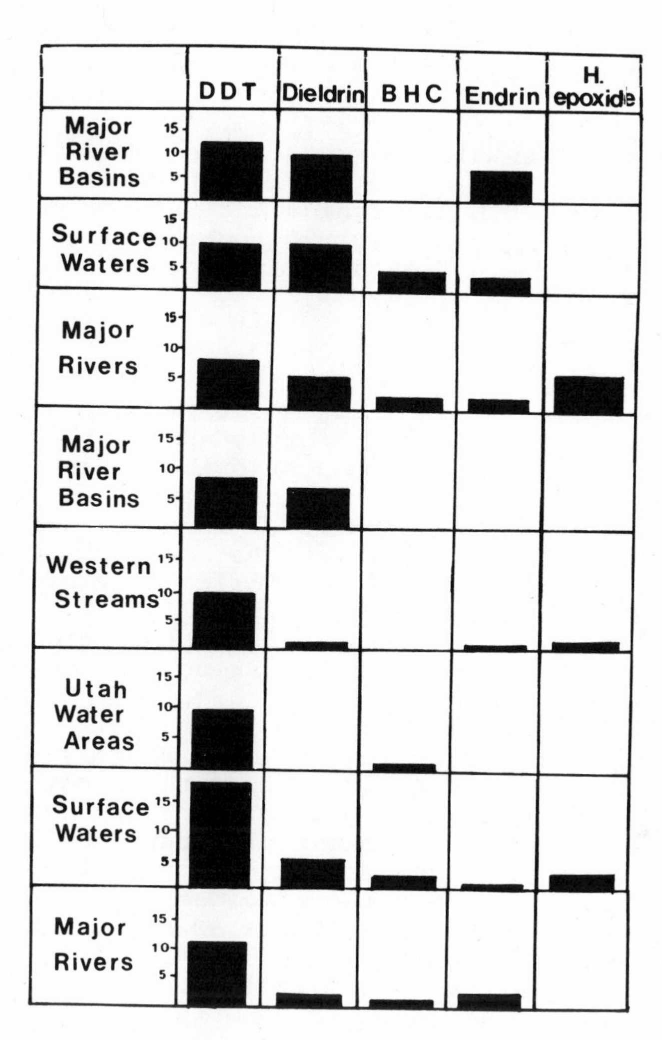

Fig. 2. Pesticide residues in U.S. waters.

LEVELS OF PESTICIDES IN AQUATIC SYSTEMS

During the last fifteen years, the monitoring of pesticide residues in various aquatic systems has increased particularly in North America. Since 1957 there has been an extensive water monitoring programme for pesticide residues in all the major waterways in U.S.A. under the supervision of a National Pesticide Water Monitoring Network. From 1962, 162 sampling sites have been monitored annually for pesticide residues (Lichtenberg et al., 1970). There have been many more local surveys in U.S.A., Canada and Europe; some of these are summarised in Fig. 2, together with some of the results of the U.S. Monitoring Network.

Rivers

Most of the available data for pesticide residues in rivers are from the U.S. Survey and nearly all the data available refer to organochlorine insecticides. Most rivers contained some residues of at least one of these insecticides; often several insecticides were reported from each site. By far the most common and largest residues have been of DDT followed by dieldrin, endrin, BHC and heptachlor epoxide (Fig. 2). DDT and dieldrin were reported from a large proportion of the rivers surveyed whereas the occurrence of the other organochlorines was much more sporadic. Camphechlor and methoxychlor have been reported occasionally but residues of chlordane are seldom found. There seemed to be a gradual increase in residues from the early 1960's reaching a peak in 1966; thereafter, there has been a gradual decline, both in terms of occurrences and residue levels. This may be the result of a gradual decline in use of organochlorine insecticides and the use of organophosphate and carbamates as substitutes; this has been accelerated by the banning of DDT in 1972 and aldrin and dieldrin in 1974. Surprisingly, there has been no increase in reports of residues of these alternative chemicals; this may be because of their relatively rapid rate of hydrolysis in water or that the analytical methods used were too insensitive to detect them.

Overall, amounts of organochlorine residues have been small, never exceeding one part per U.S. billion (1 pp 10^9) and mostly in the range of 1 - 100 mg per litre or less. Such levels are unlikely to cause any acute toxicity to fish because the L.C.-50 of the most toxic organochorine insecticide, endrin, to the most susceptible fish tested, was one part per U.S. billion.

Levels of pesticides in western streams have also been monitored in the U.S. but the amounts of pesticides reported were small ranging from 5 pp 10^{12} of lindane to 110 pp 10^{6} of DDT (Brown and Nishioka, 1967). In a later survey residues of 2,4-D up to 0.35 pp 10^{9} were reported (Manigold and Schulze, 1969).

Lakes

The pollution of lakes by pesticides seems to depend mainly upon the size of the lake and whether there is discharge of sewage or industrial effluents into them. The Great Lakes, which are surrounded by industrial complexes tend to contain appreciable quantities of persistent pesticide residues. By contrast, lakes in tropical areas remain uncontaminated by pesticides unless treated to control insects or snails. It has been suggested that the availability and toxicity of pesticide residues in lakes is correlated with the abundance and distribution of suspended particulate matter.

The Great Lakes seem to be quite heavily contaminated with DDT; in 1970 it was reported that all mature trout and salmon from L. Michigan contained DDT residues in excess of 5 ppm (Poff and Degurse, 1970). Ranges of DDT residues in fish from other Great Lakes were 0.15 - 7.77 in L. Superior, 0.65 - 6.90 in L. Huron, 0.21 - 1.89 in L. Erie and 0.40 - 4.32 in L. Ontario. In Utah Lake which is surrounded by farm land, there are seasonal peaks of pesticide residues of 1 part per U.S. billion or more of aldrin, heptachlor and BHC (Bradshaw et al., 1972).

Estuaries

Shallow coastal waters and estuaries are probably amongst the most productive of aquatic systems in terms of fish and edible crustaceans. Since estuaries receive the outflow of large rivers, much of the suspended particulate matter carried in the water is deposited in them. Some of this must be contaminated with organochlorine pesticides carried down the rivers. Because of this there has been a systematic monitoring programme of pesticide residues in estuarine molluscs at about 175 stations along the U.S. coast. These invertebrates were chosen as assay organisms because of their ability to concentrate pesticide residues from the environment. This programme has reported residues of less than 1 ppm of DDT in molluscs and only in one area were they as high as 0.1 to 0.5 ppm (Butler, 1968). Such residues are certainly too small to harm oysters and unlikely to harm fish or crustaceans that feed upon them even if biological magnification occurred.

However, another source of organochlorine residues in estuaries, is the spraying of these areas to control pests such as flies, gnats and mosquitoes. About 2 million acres of mud flats and tide areas are sprayed in this way in the U.S.; in the past mainly with DDT. Fortunately, even at these sites residues in molluscs such as oysters do not usually rise above 1 ppm (Butler, 1968). The main potential hazard is that large residues of persistent pesticides still remain in the mud of these estuaries.

Oceans

Although various workers have predicted that there has been massive contamination of the ocenas through atmospheric transport of pesticides (Riseborough et al., 1968; Goldberg et al., 1971), there is little evidence that residues in the oceans are measurable. Although, there may be contamination of these large surface areas by precipitation there is also loss from the surface through volatilisation and probably the residues in sea water fall into the ocean abyss where they may be permanently held. It has been suggested that most of the residues in the oceans are concentrated in surface water films or slicks that contain free fatty acids, fatty acid esters, fatty alcohols, and other hydrocarbons, and in plankton floating close to the surface. In one set of samples taken from the Sargasso Sea, south of Bermuda DDT residues ranged up to 2 μg/l (pp 10^{12}).

REFERENCES

Bradshaw, J.S., Loveridge, E.L., Rippee, K.P., Peterson, J.L., White, D.A., Barton, J.R. and Fuhriman, D.K. (1972) Seasonal variations in residues of chlorinated hydrocarbon pesticides in the water of the Utah Lake drainage system 1970 - 1971. Pest. Mon. J. 6, 166-170.

Brown, E. and Nishioka, Y.A. (1967) Pesticides in Selected Western Streams. A contribution to the National Program. Pest. Mon. J. 1 (2), 38-41.

Butler, P.A. (1968) Pesticides in the Estuary. Proc. March and Estuary Manage. Symp. 120-124.

Carson, Rachel (1963) Silent Spring. Houghton-Mifflin Co. 368 pp.

Chadwick, G.G. and Brocksen, R.W. (1969) Accumulation of dieldrin by fish food organisms. J. Wildlife Manage. 33, 693-700.

Edwards, C.A. (1966) Insecticide Residues in Soils. Residue Reviews 13, 83-132.

Edwards, C.A., Thompson, A.R., Beynon, K.I. and Edwards, M.J. (1970) Movement of dieldrin through soils. 1. From arable soils into ponds. Pestic. Sci. 1, 169-173.

Eichelberger, J.W. and Lichtenberg, J.J. (1971) Persistence of Pesticides in River Water. Env. Sci. Technol. 5, 541-4.

Faust, S.D. (1964) Pollution of the water environment by organic pesticides. Clinical Pharmacology and Therapeutics 5 (6), 677-686.

Goldberg, E.D., Butler, P., Meire, P., Menzel, D., Paulik, G., Riseborough, R. and Stickel, L.F. (1971) Chlorinated hydrocarbons in the marine environment. Rept. Natl. Acad. Sci. Washington D.C.

Greve, P.A. (1972) Potentially hazardous substances in surface waters. Sci. Total Env. 1, 173-180.

Hartung, R. (1972) Accumulation of chemicals in the hydrosphere. In: Environmental Dynamics of Pesticides. Ed. R. Hague and V.H. Freed. Plenum Press N.Y. and London p. 185-198.

Hill, D.W. and McCarty, P.C. (1967) Anaerobic degradation of selected hydrocarbon pesticides. J. Water Poll. Cont. Fed. 39, 1259-1277.

Hindin, E., May, D.S. and Dunstan, G.H. (1966) Distribution of insecticides sparyed by airplane on an irrigated corn plot. In: Organic Pesticides in Environment. Amer. Chem. Soc. Adv. Chem. Ser. 60, 132-145.

Kenaga, E.E. (1972) Partitioning and uptake of pesticides in biological systems. In Environmental Dynamics of Pesticides. Ed. R. Hague and V.H. Freed. Plenum Press N.Y. and London, 217-273.

Lichtenberg, J.J., Eichelberger, J.W., Dressman, R. C. and Longbottom, J.E. (1970) Pesticides in surface waters of the United States - a five year summary 1964 - 1968 Pest Mon. J. 4 (2), 71.

Lowden, G.F., Saunders, C.L. and Edwards, R.W. (1969) Organochlorine insecticides in water (Pt. 11). Water Treat. Exam. 18, 275.

Macek, K.J. (1970) Biological magnification of pesticide residues in food chains. In: The Biological Impact of Pesticides in the Environment. Ed. Gillett, J.W., Env. Hlth. Ser. No. 1. Oregon State University, 17-21.

Manigold, D.B. and Schulze, J.A. (1969) Pesticides in Selected Western Streams. A Progress Report. Pest. Mon. J. 3 (2) 124-135.

Moriarty, F. (1973) Pesticides: Significance and Implications of Biological Accumulation. AGPP: MISC./10 F.A.O. WS/E4902 14 pp.

Muirhead-Thompson, R.C. (1971) Pesticides and Freshwater Fauna. Academic Press, London and New York. 248 pp.

Nicholson, H.P. (1969) Occurrence and Significance of pesticide residues in water. J. Wash. Acad. Sci. 59, (4-5), 77-85.

Poff, R.J. and Degurse, P.E. (1970) Survey of Pesticides Residues in Great Lakes Fish. Wis. Dept. of Nat. Res. Manage. Dept. 34, 22.

Premdas, F. 11 and Anderson, J.M. (1962) The uptake and detoxification of C^{14} labelled DDT in Atlantic salmon. J. Fish Res. Bd. Canada 30, 837.

Reinert, R.E. (1972) The accumulation of dieldrin in an alga (Scenedesmus obliquus), Daphnia magna, and the guppy (Poecilia reticulata). J. Fish Res. Bd. Can. 29, 1413-1418.

Riseborough, R.W., Huggett, R.J., Griffin, J.J. and Goldberg, E.D. (1968) Pesticides: transatlantic movements in the north-east trades. Science, N.Y. 159, 1233.

Rosenberg, D.M. (1975) Food Chain Concentration of Chlorinated Hydrocarbon Pesticides in Invertebrate Communities: A Re-evaluation. Quaest Ent. 11 (1), 97-110.

Terriere, L.C., Kiigemagi, U., Zwick, R.A. and Westigard, P.H. (1966) Persistence of pesticides in orchards and orchard soils. In: Organic Pesticides in the Environment. Amer. Chem. Soc. Adv. Chem. Ser. 60, 263-270.

Thompson, A.R., Edwards, C.A., Edwards, M.J. and Beynon, K.I. (1970) Movement of Dieldrin in Soils. 11. In Soil Columns and Troughs. Pesticide Science 1, 174-178.

DYNAMICS OF PESTICIDES IN AQUATIC ENVIRONMENTS

R. Haque, P. C. Kearney, and V. H. Freed

ABSTRACT

The water solubility, adsorption, and bioaccumulation of pesticides and related chemicals have been discussed in the light of dynamics of pesticides in the aquatic environments. The solubility of chlorinated biphenyl analogs increased with a decreasing number of chlorine atoms in the molecule. In contrast, adsorption of polychlorinated biphenyls on clay, sand, soils, and humic acid increased as the number of chlorine atoms increased. The octanol/water partition coefficient of organophosphate pesticides varies over a wide range. Ethoxy groups usually increase the partition coefficient, as compared with the corresponding methoxy analogue. The bioaccumulation of various pesticides in fish was influenced by polarity and water solubility. There is a qualitative correlation between water solubility and fish bioaccumulation--the bioaccumulation decreased as the solubility of the pesticide increased. Water solubility is important in decreasing the dynamics of pesticides in aquatic environment.

INTRODUCTION

Once a pesticide is introduced into the environment, there is a reasonable chance that it will eventually find its way into water. Therefore, aqueous systems probably represent one of the most important complex environments as far as describing the fate and behavior of pesticides.

Many factors influence the dynamics of a pesticide in the aquatic environment, including: water solubility; hydrolysis; complex formation; adsorption to soils and sediments; partitioning into lipids; and bioaccumulation in living organisms. In this paper, we describe some of these factors in relation to the study of the dynamics of pesticides in the aquatic environment. No attempt has been made to review the literature, but only pertinent findings, mostly from the work of the authors will be discussed.

Water Solubility

The water solubility of a pesticide provides information on the extent to which the pesticide will partition in an aquatic environment. The salts and organic matter in water, and variation in pH and temperature will affect their water solubility. Pesticides range widely in their water solubilities (Table I). For example, diquat and paraquat are highly water soluble, whereas DDT has a water solubility of 1.2 parts per billion (ppb). In general, the solubility of most pesticides ranges in parts per million (ppm). The determination of water solubility of pesticides with values in the ppm range or more is quite easy. However, for strongly hydrophobic pesticides, like DDT whose solubility is in the ppb range, the determination of water solubility is difficult because such chemicals form aggregates in aqueous solution. Recently, Haque and Schmedding (1975) developed a method for the measurement of solubility of hydrophobic chemicals. This method involves a slow equilibration of the chemical in water. The chemical under investigation is first deposited as a thin film inside the wall of a large vessel, like a carboy; water is added; the system is slowly stirred, and the concentration of the chemical is determined at various intervals of time. With several polychlorinated biphenyls, up to 6 months might be needed to reach equilibration. Continuous stirring may cause the formation of aggregates and give higher water solubility values. To obtain the true solubility, water concentration should be determined after stirring has stopped (Fig. 1). The solubility of various polychlorinated biphenyls are given in Table II. The solubility values reported by Wallnofer et al. (1973) were consistently higher than those reported by Haque and Schmedding (1975). This is the obvious effect of the aggregate formation. As expected, the solubility decreases as the number of chlorine atoms in the molecule increased.

Table I. Water Solubility of Some Pesticides

Pesticide	Solubility ppm	Pesticide	Solubility ppm
Atrazine	70	p,p'-DDT	.0012
Monuron	230	Lindane	10
2,4-D	700	PCP	20
Diquat	very soluble	Parathion	24
Amitrole	very soluble	Leptrophos	.0047

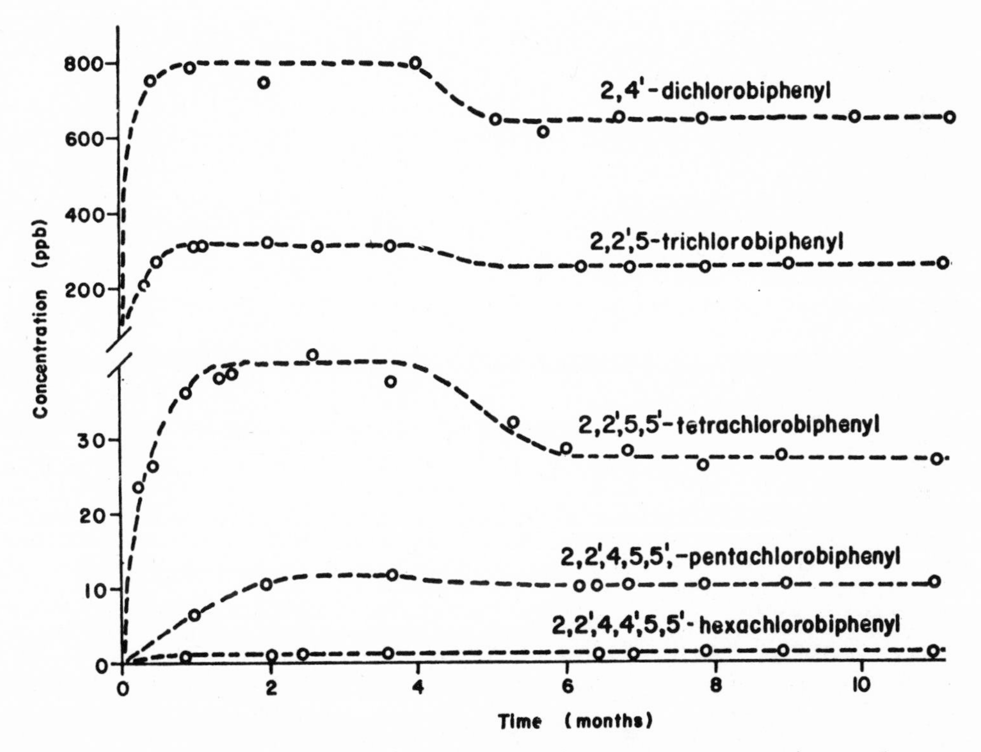

Figure 1. The solubility of various polychlorinated biphenyls in water as a function of time. From Haque and Schmedding (1975).

Table II. Water Solubility of Five PCB Isomers

	Solubility (ppb)	
	Wallnofer et al. (1973)	Haque and Schmedding (1975)
2,4'-Dichloro	1850-1900	637 ± 7
2,2',5-Trichloro		248 ± 4
2,2',5,5'-Tetrachloro	46	26.5 ± 0.8
2,2',4,5,5'-Pentachloro	28-35	10.3 ± .2
2,2',4,4'5,5'-Hexachloro	8.8	.953 ± .01

Adsorption

Adsorption of pesticides to soils and sediments is one mechanism through which a pesticide is removed from the aquatic environment. Adsorption is a complex process that depends upon the nature of organic impurities and salts in water. The transport of hydrophobic chemicals in the environment is related to their adsorption to particulate matter. Some data on the adsorption of a few polychlorinated biphenyls on soils, clays, and humic acid have been reported (Haque and Schmedding, 1976). Adsorption of chlorinated biphenyls on Woodburn soil, illite clay, humic acid, and Del Monte Sand is shown in Fig. 2. Adsorption increases qualitatively as the number of chlorine atoms in the molecule increases. The data are represented by the Freundlich-type isotherm (i) where: χ is the amount of pesticide; m is

$$\frac{\chi}{m} = KC^n \qquad (i)$$

the mass of the adsorbent; C the equilibrium concentration of the pesticide, and K and n are constants. These constants, together with the water solubility, are given in Table III. In general, the adsorption follows the series humic acid > illite clay > Del Monte sand. The Freundlich constant, K, which is an indirect measure of the extent of adsorption, increased as the number of chlorine atoms increased in the molecule. The K values for humic acid were unusually high, which may be due to the high surface area of humic acid and various functional groups on this surface. The constant n, which deals with the nature of the adsorption, was in the range of unity, except for the tetrachloro- and hexa-

Table III. Adsorption Characteristics and Water Solubility of PCB Isomers

Adsorbate	Water Solubility (ppb)	Adsorption Characteristics					
		Illite Clay		Woodburn Soil		Humic Acid	
		n	K	n	K	n	K
2,4'-dichlorobiphenyl	637 ± 7	.84	20.0	1.18	7.00	.86	3.98 x 10
2,5,2',5'-tetrachloro-biphenyl	26 ± .8	.92	21.2	1.00	48.9	3.82	4.36 x 10
2,4,5,2',4',5'-hexachloro-biphenyl	.953 ± .01	1.26	200	1.25	320	3.78	3.39 x 10

chloro- compounds on humic acid. The high n value represents an S-type adsorption, where the solvent molecules were not appreciably adsorbed on the surface. Humic acid, being strongly hydrophobic, was probably a poor adsorber for water. As expected, the water solubility reflects an inverse correlation with the K values. This also implies that the crystallization of the adsorbate on the surface probably was involved in the mechanism of adsorption. The desorption of these chemicals was negligible (4%) at the highest adsorption level. In view of the low desorption and high adsorption of these chemicals, their transport may be attributed via adsorption on waterborne particulates. This supports the findings of Duke et al. (1966), who noted polychlorinated biphenyls concentrations several times higher than the aqueous solubility limit.

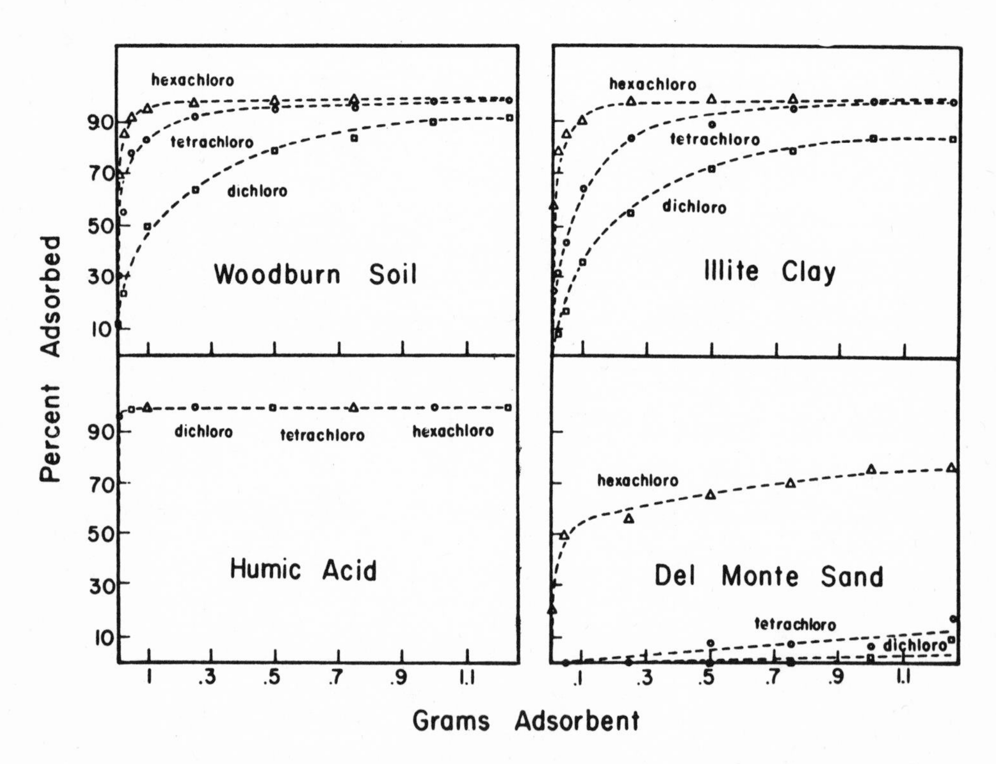

Figure 2. Adsorption of various polychlorinated biphenyls on Woodward soil, Illite clay, humic acid and Del Monte sand. From Haque and Schmedding (1976).

Bioaccumulation

The bioaccumulation of pesticides in aquatic organisms defines the potential of pesticide transport in the food chain. Bioaccumulation depends upon the physical chemical characteristics of the pesticide, environmental conditions, and the nature of the aquatic organism. The bioaccumulation of pesticides in aquatic organisms encompasses a large range of values; organochlorine-type compounds, being more hydrophobic, bioaccumulate more than do most other pesticides. Bioaccumulation can be measured directly or indirectly. The direct method involves a model ecosystem study, where the concentration of the pesticide is determined among various aquatic organisms. The indirect method involves the measurement of octanol/water partition coefficient of the pesticide. From the magnitude of the partition coefficient data, the extent of bioaccumulation of pesticides can be predicted qualitatively. Such a prediction was based on the correlation which existed between partition coefficient and bioaccumulation potential of many organic chemicals (Fig. 3) (Neeley et al., 1974). The two methods in predicting bioaccumulation potential are described as follows:

<u>Indirect Method; Partition Coefficient:</u> The partition coefficient P used in predicting bioaccumulation can be defined according to the equation

$$P = \frac{[C_1] \text{ octanol}}{[C_2] \text{ water}} \qquad \text{(ii)}$$

where C_1 and C_2 are the concentrations of the pesticide in octanol and water, respectively. Although in theory the determination of P is quite straightforward, actually several experimental problems are encountered. A major problem is the finite solubility of octanol in water, which facilitates the formation of small aggregates of octanol in water. Recently, Freed et al. (1976) measured P values for a number of organophosphate insecticides. With this method both octanol and water are saturated with each other before starting the experiment. A known amount of pesticide in octanol was shaken with water in a glass centrifuge tube in a horizontal position for 24 hours at 20°C, and the concentration of the pesticide was determined in octanol and water, respectively. The P values of several of

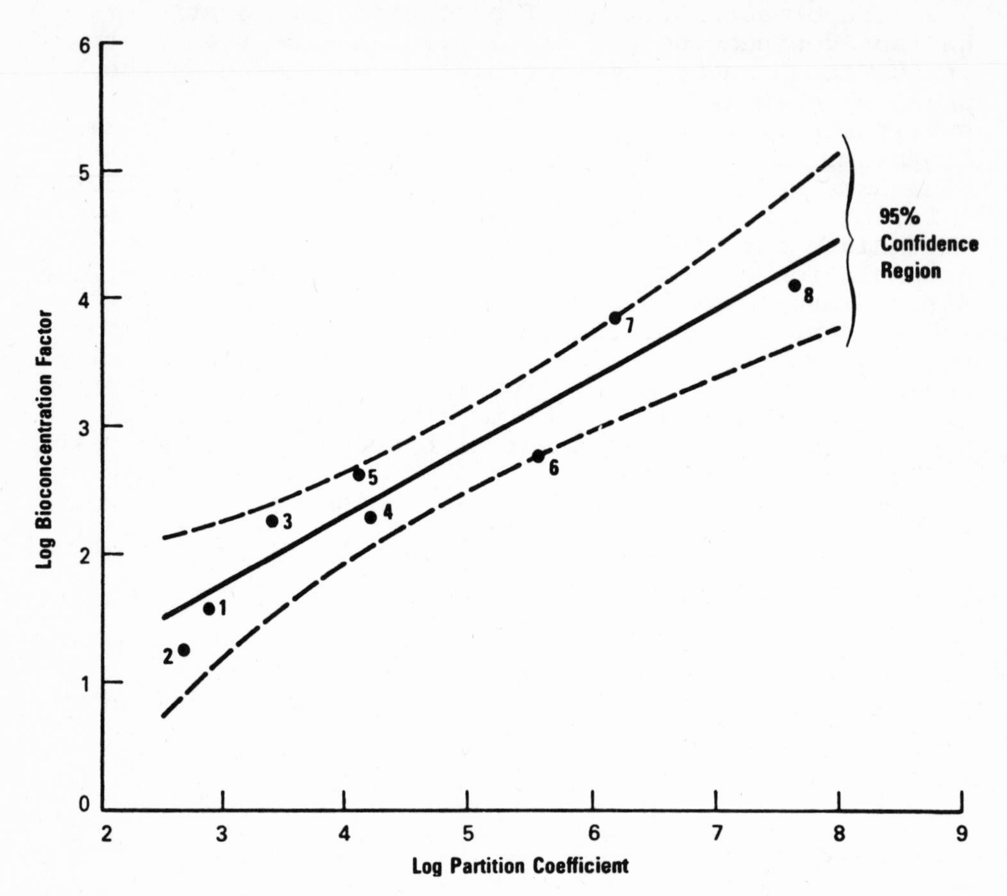

Figure 3. The linear regression between the logarithms of the partition coefficient and the bioconcentration of various chemicals in trout muscle. (The numbers refer to the following chemicals:

1. 1,1,2,2-tetrachloroethylene
2. carbon tetrachloride
3. p-dichlorobenzene
4. diphenyl oxide
5. diphenyl
6. 2-biphenyl phenyl ether
7. hexachlorobenzene
8. 2,2'4,4'-tetrachlorodiphenyl)

these organophosphates are given in Table IV. An examination of these data suggested that organophosphates possess P values ranging from 5×10^2 to 2×10^6. The unusually high P values for leptophos may be attributed to its unique structural characteristics and the presence of bromine in the molecule.

Table IV. Partition Coefficient of Several Organophosphates

Compound	Partition Coefficient
Parathion	6,430
Dicapthon	3,790
Fenitrothion	2,380
Dichlofenthion	137,000
Ronnel	75,300
Methyl Chloropyrifos	20,000
Chloropyrifos	128,000
Leptophos	2.02×10^6
Phosmet	677
Dialifor	49,300
Phosalone	20,000
Malathion	781
Dimethoate	508

Some correlation between P values and structure of organophosphates can also be drawn. In compounds possessing similar structures, the replacement of an ethoxy group with methoxy results in a decrease in the P values. This was supported by the P values of dialifor vs. phosalone, and chloropyrifos vs. methyl chloropyrifos, and to some extent parathion vs. dicapthon. Compounds, like malathion and dimethoate possessing no benzene ring, have very low P values.

Direct Method: Model Ecosystem Studies: A unique concept of studying the distribution of pesticides in various organisms, commonly known as model ecosystem, was introduced by Metcalf et al. (1971). This approach provides a good indication of how a chemical may accumulate in various organisms. One of the main advantages of the model ecosystem was that experiments were carried out in a relatively simple way under controlled conditions. The classical model ecosystem of Metcalf et al.

(1971) included a glass aquarium 10" x 12" x 20" in dimensions with 180 square inches of surface area. Quartz sand with a sloping profile was placed and standard reference water was added. The aquarium contained organisms, like sorghum, algae, plankton, snails, cladocera, mosquito larvae, and fish. The main parameter used in model ecosystem studies was bioaccumulation or biomagnification. Bioaccumulation was associated with the accumulation of the chemical in the organism through absorption, adsorption and ingestion, whereas biomagnification was related to the concentration of the chemical by consumption of food of lower by higher food chain organisms. Bioaccumulation ratio was calculated using the following relationship:

$$\text{Bioaccumulation Ratio} = \frac{\text{Concentration of chemical in organism}}{\text{Concentration of chemical in water}} \qquad \text{(iii)}$$

Earlier studies by Metcalf et al. (1971) showed that DDT was bioaccumulated about 100,000 fold and DDE to 30,000-50,000 in fish. Model ecosystems have been developed by many other workers for predicting bioaccumulation potential for other pesticides. We will describe some of the results from the laboratories of Isensee et al. at Beltsville, Maryland. This model ecosystem is a modification of the one used by Metcalf et al. (1971), and in the experiment control and ^{14}C-labeled pesticide treated soil was placed in glass tanks containing 4 liters of water. A day later, 100 daphnids (*Daphnia magna*), eight snails (*Physa* sp.), a few strands of an alga (*Oedogonium cardiacum*), and 10 ml of old aquarium water, containing various diatoms, protozoa, and rotifers, were added. Water samples were analyzed at 2- day intervals, and at 30 days, samples of daphnids were taken out for analysis and two mosquito fish (*Gambusia affinis*) were added to each tank. After three days, all the organisms were analyzed for the pesticides concentration. A recently developed, larger model-ecosystem in which fish were exposed for 30 days is shown in Fig. 4. The results of such studies are summarized in Table V. It is apparent from the table that a bioaccumulation ratio for the same chemical depended upon the concentration of the pesticide, as well as other environmental conditions. Although bioaccumulation ratios varied, it is interesting to note that solubility of the chemical has a profound effect on it. On a qualitative basis, chemicals possessing

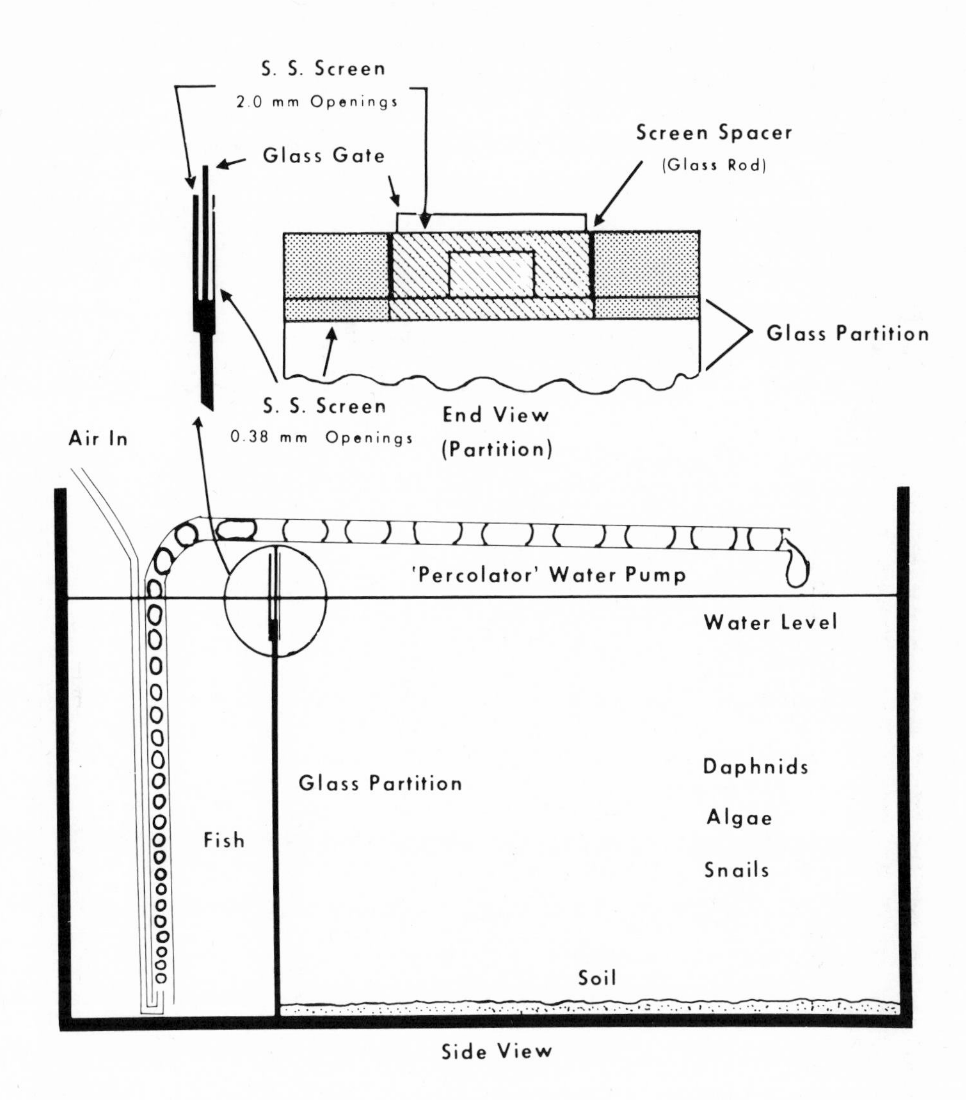

Figure 4. Recirculating static model ecosystem. Tank dimensions: 41x20x24 cm; glass partition 18 cm high. Tank volume 16 liters with a .1 cm water depth over the glass partition. From Isensee (1976).

Table V. Bioaccumulation ratio (BR) and water solubility of five compounds in various aquatic organisms. From Isnee (1976), Isnee and Jones (1975) and Kanazawa et al. (1975).

Chemical	Water Solubility	Organisms	BR
Carbaryl	40 ppm at 30°C	algae	4000
		duckweed	3600
		snails	300
		catfish	140
		crayfish	260
Atrazine	33 ppm at 25°C	snails	2-15
		algae	10-83
		fish	3-10
HCB	practically insoluble	algae	320-1570
		snails	1360-3320
		daphnids	770-1030
		fish	1160-3740
Mirex	insoluble	algae	12,200
		snails	4,900
		daphnids	14,650
		fish	2,580
TCDD	.006 ppm 30°C	algae	2000-18,600
		duckweed	1200-5000
		snails	1400-47,100

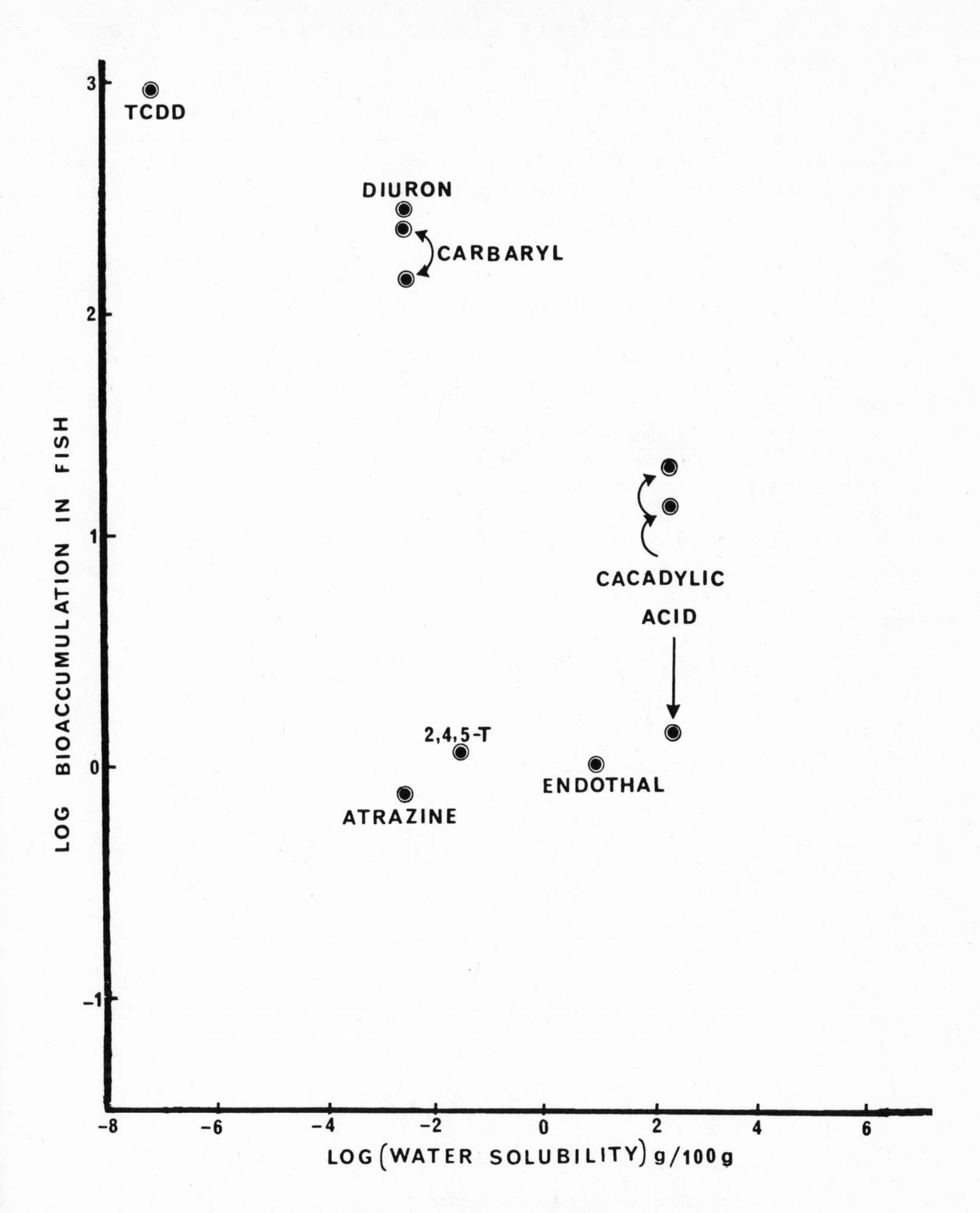

Figure 5. Plot of log water solubility and log bioaccumulation in fish for eight organic pesticides.

high solubility have lower bioaccumulation. This is more apparent in Fig. 5.

Solubility, then, offers one of the simplest parameters for predicting the adsorption characteristics of a pesticide in the aquatic environment. Once accurately measured, it provides a reasonable indicator for potential bioaccumulation.

LITERATURE CITED

Freed, V. H., Haque, R., Schmedding, D. and Koluist, R. 1976. Environ. Health Persp., 13: 77.

Duke, T. W., Lowe, I. J. and Wilson, A. J. Jr. 1966. Bull. Environ. Contam. Toxicol. 5: 171.

Haque, R. and Schmedding, D. 1975. Bull. Environ. Contam. Toxicol. 14: 13.

Haque, R. and Schmedding, D., J. 1976. Environ. Sci. Health, B11 (2): 129.

Isensee, A. R. 1976. Int. J. Environ. Studies 9, accepted for publication.

Isensee, A. R. and Jones, G. E. 1975. J. Environ. Sci. Tech. 9: 668.

Kanazawa, J., Isensee, A. R. and Kearney, P. C. 1975. J. Agr. Food Chem. 23: 760.

Metcalf, R. L., Sangha, G. K. and Kapoor, I. P. 1971. Environ. Sci. Tech. 5: 709.

Neely, W. B., Bronson, D. R. and Blau, G. E. 1974. Environ. Sci. Tech. 8: 1113.

Wallnofer, P. R., Koniger, N. and Hutzinger, O. 1973. Analab Res. Notes 13: 14.

FATE OF PESTICIDES IN AQUATIC ENVIRONMENTS

H. V. Morley

INTRODUCTION

In the short time available it is obvious justice cannot be done to the subject of this presentation and that a high degree of selectivity combined with a great deal of ability to compress and summarize data will be required. For this reason it will be assumed that the pesticide has already been transported to aquatic ecosystems from direct application, from runoff waters, or from the atmosphere. Once it is in the aquatic system, it may be rapidly degraded by a variety of mechanisms or persist for a period of time. Its persistence and, therefore, its potential for biomagnification will be determined by a variety of factors. Perhaps the most important single factor is the ability to partition into lipid-rich pools. Fat-soluble, water-insoluble materials, e.g., DDT, PCBs, accumulate into "environmental sinks" and, by virtue of their water insolubility, resist the usual transformation and detoxification mechanisms and provide a virtually constant input source of material. These are the environmental "bad guys" and one must be careful not to include all pesticides in this category. Very few of these persistent lipid-soluble pesticides remain on the market today. Their replacements are, relatively speaking, more water soluble and will partition between lipid-rich pools such as biota, suspended solids, sediments and water. Monitoring studies indicate that there is little, if any, buildup of the newer materials in aquatic ecosystems and two Canadian field studies will be presented to support this thesis. It may be argued, however, that it is too soon to make such a statement and that given the time and volume of usage that DDT had, we may come up with new candidate "bad guys". This is doubtful, however, since the physical and chemical properties of materials in use nowadays would seem to preclude any such long term persistence and biomagnification potential. The ability of microorganisms to break down "solubilized" materials seems to know no limits.

First one must define what is meant by the term "fate". One of the dictionary definitions is "what is destined to happen" and this is as good a definition as any for our purposes. What is destined to happen to pesticides in an aquatic environment? To answer this question in full will perhaps never be possible when the complexity and variability of the system under investigation is considered. This very complexity and variability of the aquatic ecosystem make it almost impossible to extrapolate laboratory experiments to the field no matter how well designed they may be. Apart from this consideration, there are two other important interrelated factors which must be considered. The first of these is the increasing skill of the pesticide residue chemist, the other is a basic lack of knowledge regarding, and of criteria for evaluating, the ecotoxicity of the pesticide and its transformation products.

Many of the questions that we are asking nowadays seem to reflect the increasing ability of the residue chemist to measure the unmeasurable! Back in the halcyon days of the late '50s and early '60s the residue chemist talked of parts per million (10^{-6}), used for the most part non-specific methodology and pesticide tolerances were commonly in the 5-10 ppm range or higher. With the advent of gas chromatography and the electron capture detector, the residue analyst easily broke through the ppm (10^{-6}) barrier and parts per billion (10^{-9}) became the "norm", and food tolerance reflected this by a downward change in levels. Now our degree of analytical sophistication is such that parts per trillion (10^{-12}) are being routinely reported. In addition, positive identification of residues is now readily achieved by many methods but especially by use of mass spectrometry. It is obvious that with the highly sensitive techniques available and being developed, we are going to find more and more materials at lower and lower levels and, moreover, all potential pollutants of the environment.

What we do not know, however, is what, if any, effect these ultra-low levels of seemingly non-persistent compounds is having on the environment. Criteria for assessment of the importance or the ecotoxicity of these materials at the levels being found in the environment is lacking. I have perhaps digressed from my subject matter but felt it necessary to point out that a detailed accounting of the fate of pesticides in the environment, while highly desirable in terms of knowledge, can have little significance until toxicologists develop adequate criteria for testing and evaluating the effect of the pesticide and/or transformation products on the ecosystem. The problem of synergistic and antagonistic effects of mixtures of chemicals and their potential interaction with viral and bacterial agents is yet another "can of worms" that some intrepid toxicologists must tackle sooner or later.

Two Canadian field studies regarding the fate of pesticides in aquatic ecosystems have been underway for the past few years. At least one more year in each of the programmes is planned but the results to date will be described. It must be emphasized that much of the data is still being processed and that any conclusions must, therefore, be provisional and treated with a certain degree of caution. The two studies in question are: (1) The fate of pesticides in representative agricultural watersheds in Ontario, and (2) The fate of methoxychlor in a river ecosystem.

THE FATE OF PESTICIDES IN AGRICULTURAL WATERSHEDS

The fate of pesticide residues in the Great Lakes region has long been of concern to Canada. An excellent review of the early work in this area has recently been published. (Harris and Miles, 1976)

Impetus and resources for more detailed studies were provided as a result of the signing of the Great Lakes Water Quality Agreement by the Governments of Canada and the United States on April 15, 1972. One of the sections of the Agreement requested the International Joint Commission (IJC) to conduct a study of pollution of the boundary waters of the Great Lakes system resulting from agricultural, forestry, and other land use activities.

One of the potential pollutants of the Great Lakes system was pesticides arising from agricultural management practices. Tackling the problem presented some difficulty since available pesticide residue monitoring data from the Great Lakes indicated that pesticides being found were primarily the organochlorine pesticides which had been largely banned and phased out in the early 1970s in both Canada and the U.S.A. Replacement materials now being used were not being found but it was argued that this might be due to the relatively short time that they had been in use rather than their relatively short persistence. Samples collected from the drainage ditches adjacent to the Bradford Marsh in Ontario indicated levels of ethion, parathion and diazinon present in the water. (Harris and Miles, 1975) These pesticides were being used for control of onion maggot and carrot flies. Levels of diazinon were particularly striking since while low at the beginning of the season, they steadily increased reaching peaks of 1.07, 0.68 and 2.04 ppb in the fall. These results were obtained, however, at a time when no ditch water was being pumped into the Schomberg River. Results in other sampling locations were not as high and considerable variation in values were obtained. Nevertheless, persistence of residues of diazinon in water appears to exceed six months. (Harris and Miles, 1976)

During the same period that saw the persistent organochlorine insecticides being replaced by less persistent organochlorines, the organophosphorus and carbamate insecticides, there was a rapid increase in the amount of herbicides being used. By 1973 herbicides had become the single largest group of pesticides used in Ontario. Thus, of the total quantity of pesticide active ingredients applied to field crops (4,509 tonnes), 58.8% (2.652 tonnes) were herbicides. (Roller, 1973)

The purpose of this study was to establish the fate of known inputs of pesticides currently being used into agricultural watersheds in an attempt to determine if a problem exists with regard to pesticide input to aquatic ecosystems. Any identifiable problems could then be addressed with emphasis on transportation and transformation of pesticides. Some of the preliminary findings will be presented here but it must be emphasized that these results are tentative and preliminary in nature as there are still many gaps in the data and many results are still being processed. Loading of pesticides is based on provisional daily discharge flow rates of the rivers.

TABLE 1.- Ontario Agricultural Sub-Watersheds Monitored for Pesticides (1)

Watershed location	County	Area (ha)	Soil type	Main agricultural practices
Big Creek	Essex	5180	Clay	Corn, soybeans
Venison Creek	Norfolk	7374	Sand	Tobacco
Little Ausable River	Huron	5410	Clay	Corn, cattle, hogs
Canagagigue Creek	Wellington	1890	Loam	Cereals, dairy, hogs
North Creek	Niagara North	2990	Clay	Corn, hogs, poultry
Hillman Creek	Essex	4080	Sand	Corn, soybeans, vegetables, fruits

(1) Data from five other agricultural sub-watersheds not yet processed.

TABLE 2.- Polychlorinated Biphenyls in Stream Water from Six Agricultural Sub-Watersheds, 1975 (Frank et al, 1976)

Watershed location	Water samples	Concentration (ppb)	
		Range	Mean
Big Creek	49	0.01-0.12	0.04
Venison Creek	29	0.01-0.18	0.08
Little Ausable River	46	0.01-0.32	0.08
Canagagigue Creek	23	0.01-0.14	0.09
North Creek	14	0.01-0.11	0.03
Hillman Creek	52	0.01-0.60	0.04
Totals and Means	213	0.01-0.60	0.07

In order to obtain information on the fate of the new pesticides being used in water and sediments, work was initiated on 11 agricultural sub-watersheds in Southern Ontario chosen so as to represent all possible soil types, climatic zones and farm practices. Surveys were carried out by questionnaire (Roller, 1973) and by farm-to-farm survey in 1974 to determine the type of crop production, and amount and type of pesticide being used. This information was used to establish total pesticide input into the selected watersheds. Analyses of integrated water samples collected at the outlets of each of the agricultural watersheds were carried out at normal flow pattern and at most rainfall event times. This work was carried out by the Ontario Ministries of Agriculture and Food, and of the Environment. Based on a knowledge of the pesticides being used, analyses were routinely carried out for over 40 pesticides and metabolites. In addition to pesticides, PCB levels were determined.

To establish if there was any additional input via the aerial route, rainfall measurements were made in each watershed (Sanderson and Osborne, 1976) and collected samples analyzed for insecticides and PCBs. (Frank et al, 1975)

Between March 1975 and January 1976, 873 samples were collected and 1,723 analyses carried out on stream water collected at monitoring sites of 11 agricultural sub-watersheds. Only the results from the watersheds listed in Table 1 are considered since data from the other five watersheds are still being processed.

Polychlorinated Biphenyls (PCBs)

Although not a pesticide, PCBs are included here since the data are of interest insofar as atmospheric input figures would appear to be high enough to account for the relatively uniform distribution and the amounts being found in the streams. Thus, PCBs were present in all water samples analyzed, at levels below 1 ppb (1 x 10^{-9}g) (Table 2). Levels in rainfall contained PCB levels from below detection limits (<0.01 ppb) to 0.10 ppb. From the amount of rain which fell between May and December an estimate was made of the total PCB fallout in the watersheds. This ranged from 184 to 277 mg/ha (Table 3). Discharge rates from the stream were calculated using stream flow data and PCB levels. For the study period water levels ranged from 0.03 ppb in

TABLE 3.- Polychlorinated Biphenyl (PCB) Fallout in Rain, May-December 1975 and Discharged in Water, April-December 1975 (Frank et al, 1976)

Watershed location	Rainfall ($10^6 m^3$)	PCB fallout		Stream discharge ($x\ 10^3 m^3$)	PCB discharged	
		(g)	mg/ha		(g)	% of fallout
Big Creek	5.18	975	188	121.0	4.8	0.5
Venison Creek	-	-	-	521.0	41.1	-
Little Ausable River	6.49	1424	263	213.0	16.8	1.2
Canagagigue Creek (1)	2.65	481	255	121.0	11.3	2.4
North Creek	3.58	549	184	79.5	2.6	0.5
Hillman Creek	5.71	1129	277	65.5	2.6	0.2

(1) Fallout May to September

North Creek to 0.02 ppb in Canagagigue. These rather uniform concentrations support the idea that atmospheric fallout could be the main input to the agricultural watersheds in Southern Ontario. These levels correspond to 0.2 to 2.4% of the aerial input being removed in stream discharge (Table 3). These figures do not, however, take into account PCB movement and levels in bottom sediment, making a mass-balance impossible at this stage.

Organochlorine (OCs) Pesticides

As mentioned previously, use of most of the OCs has been virtually phased out since 1970 and presence in the watersheds reflect past use. Nevertheless, DDE was present in 92% and dieldrin in 27% of the water samples analyzed. The parent material DDT was found in only 10% and TDE in 24% of the samples (Table 4).

Of particular interest was the aerial input of DDE which ranged from 14-31 mg/ha or 27 to 127 g per watershed (Table 4). Discharge figures are also given, but again, movement and levels in bottom sediment are not available. Nevertheless, discharge from Venison Creek can be seen to be much higher than the other five watersheds reflecting the large quantities of DDT used in the past in this tobacco growing area. Use of DDT did not stop until two years after the other watersheds, i.e. in 1972.

Endosulfan was a major replacement for DDT. The agricultural survey indicated use in only three watersheds, but residues were detected in water from all six watersheds (Table 5). The low levels in the Little Ausable and the Canagagigue could be accounted for by domestic use. The relatively high levels found in North Creek prompted a closer review of all data but no commercial crops could be found grown in 1975 for which endosulfan had a registered use. There were, however, many small gardens close to the stream which might account for its presence in 14% of the samples examined. The majority of the samples comprised mainly the sulfate metabolite.

Chlordane was a major replacement for aldrin and dieldrin since use of these materials was banned in Ontario in 1969. Concern has been expressed at the fate of the heptachlor impurity present in chlordane formulations and the possible impact of chlordane on aquatic ecosystems. (NRC, 1973) Unexplained is the

TABLE 4.- Fallout of DDE in Rain, May to December 1975, and Discharge of DDE, TDE and DDT in Water, April to December 1975 (Frank et al, 1976)

Watershed location	Fallout DDE		Discharge water		Presence in water (%)			Concentration (x 10^{-12}g)		
	(g)	(mg/ha)	DDE (g)	TDE & DDT (g)	DDE	TDE	DDT	DDE	TDE	DDT
Big Creek	77.1	15	1.27	0.16	88	22	12	10.5	0.9	0.4
Venison Creek	NP (2)	NP	4.95	6.71	97	38	21	9.5	1.7	11.2
Little Ausable River	109.0	20	3.03	0.24	91	13	2	14.3	1.1	<0.1
Canagagigue Creek (1)	27.2	14	2.30	0.03	83	13	4	19.4	0.2	<0.1
North Creek	77.1	26	0.21	0.02	86	7	0	2.7	0.3	ND
Hillman Creek	127.0	31	0.20	0.20	98	37	15	14.1	0.7	2.4

(1) Fallout May to September only
(2) NP = Data not yet processed

TABLE 5.- Endosulfan Inputs and Outputs in Six Agricultural Sub-Watersheds in Ontario, 1975 (Frank et al, 1976)

Watershed location	Agricultural input (kg)	Stream output (g)	Presence in water (%)	Mean concentration ($x10^{-12}$)
Big Creek	11	0.54	22.0	4.5
Venison Creek	519	1.76	28.0	3.4
Little Ausable River (1)	-	0.07	6.5	0.4
Canagagigue Creek (1)	-	0.01	4.4	0.1
North Creek (1)	-	0.27	14.0	3.4
Hillman Creek	70	0.85	52.0	13.0

(1) Suspected domestic uses in the watershed. No crops grown in North Creek to warrant use of endosulfan.

observation that in 1975 the only documented use of chlordane was in the Little Ausable River watershed and yet no heptachlor epoxide was detected in the water (Table 6). On the other hand, heptachlor epoxide was found in samples from both the North and Hillman Creeks where there was no recorded use. The small levels found could possibly be explained by past years' usage of either heptachlor or chlordane.

Organophosphorus Insecticides (OP)

Twelve OPs were used in the six watersheds. The four major materials used were azinphos methyl, chlorpyriphos, phosmet and leptophos (Table 7). In spite of the relatively large uses, residues of these materials could not be detected except for one sample (out of 30) in Venison Creek which contained chlorpyrifos at 1×10^{-2}g level. On the other hand despite the relatively low input of diazinon, residues were detected in 35% of 51 samples from Hillman Creek at a mean concentration of 2.3×10^{-9}g. The limited commerical input does not preclude the possibility of non-agricultural use and this possibility is being further examined. This finding would seem to support the findings of Harris and Miles concerning the persistence of diazinon residues in aquatic ecosystems.

Herbicides - Atrazine

In all six watersheds corn is a major crop ranging between 10% in Venison Creek to 32% in the Little Ausable River. On an average 21% of the watersheds were devoted to corn and 70% of the corn was treated with atrazine. Inputs of atrazine are given in Table 8, ranging from 497 to 1,951 kg per watershed.

In four of the six watersheds all water samples contained measurable residues of atrazine (Table 9). The deethylated atrazine metabolite was present in 33% to 93% of the samples from the watersheds.

The peak concentration of atrazine in water occurred in June when a mean concentration of 5 ppm was reached. This declined to 2 ppb in July and was at or below 0.5 ppb for the remainder of the year. In one watershed almost 95% of the annual loss occurred in one major rainstorm in June.

TABLE 6.- Heptachlor Epoxide in Discharge Water from Present Chlordane Use or Past Uses of Heptachlor (Frank et al, 1976)

Watershed location	Input of chlordane (kg)	Output (g)	Presence in water (%)	Mean residue ($x10^{-12}g$)
Little Ausable River	1041	ND	ND	ND
North Creek	-	0.01	7.1	0.1
Hillman Creek	-	0.32	11.5	4.8
Total or Mean	1041	0.33	3.3	0.3

TABLE 7.- Organophosphorus Insecticide Input and Output in Two Ontario Agricultural Sub-Watersheds (Frank et al, 1976)

Watershed location	Insecticide input (kg)		River Output (g)	Presence in water	
				No. of Samples (%)	Mean Concentration ppb
Venison Creek	chlorfenvinphos	(72)	ND	-	-
	chlorpyriphos	(1423)	0.68	3.5	0.001
	leptophos	(568)	ND	-	-
	malathion	(7)	ND	-	-
	trichlorfon	(373)	ND	-	-
Hillman Creek	azinphos methyl	(2420)	ND	-	-
	chlorpyriphos	(7)	ND	-	-
	demeton	(3)	ND	-	-
	diazinon	(3.5)	150.0	35.0	2.3
	dimethoate	(2)	ND	-	-
	leptophos	(34)	ND	-	-
	malathion	(11)	ND	-	-
	phosalone	(213)	ND	-	-
	phosmet	(664)	ND	-	-

TABLE 8.- Atrazine Inputs and Outputs in Six Ontario Agricultural Sub-Watersheds, 1975 (Frank et al, 1976)

Watershed location	Inputs (kg)	Output		Ratio	
		Atrazine (g)	de-ethyl-atrazine (g)	Out/In (%)	Parent/metabolite
Big Creek	724	287	24	0.043	12.0
Venison Creek	1289	89	68	0.012	1.3
Little Ausable River	1951	284	108	0.020	2.6
Canagagigue Creek	932	87	27	0.012	3.2
North Creek	497	115	12	0.026	9.6
Hillman Creek	595	12	3	0.003	4.0

TABLE 9.- Atrazine and its Metabolite in Stream Water of Six Ontario Agricultural Sub-Watersheds (Frank et al, 1976)

Watershed location	Presence in water (%)		Atrazine concentration in water (ppb)		de-ethyl-atrazine concentration in water (ppb)	
	Atrazine	de-ethyl-atrazine	Mean	Range	Mean	Range
Big Creek	100	45	2.4	< 0.1-17.7	0.2	ND-3.7
Venison Creek	41	45	0.2	ND-08.3	0.1	ND-4.8
Little Ausable River	100	93	1.3	0.2-31.7	0.5	ND-2.5
Canagagigue Creek	100	91	0.7	< 0.1-12.1	0.2	ND-2.6
North Creek	100	57	1.4	0.1-09.6	0.2	ND-1.1
Hillman Creek	60	33	0.2	ND-04.4	< 0.05	ND-1.0

Conclusion

It is always dangerous to make conclusions from an incomplete study but it would appear to be safe to say that present-day pesticides appear to pose few problems with regard to their potential for biomagnification and long term damage to the aquatic ecosystem. Of the pesticides examined, only atrazine and diazinon appear to occur in streams with any regularity, albeit at the ppb level. The ecotoxicity, if any, of ultralow levels of pesticides in the environment remains to be determined in most cases.

THE FATE OF METHOXYCHLOR IN A RIVER ECOSYSTEM

Sporadic, severe outbreaks of a black fly, *Simulium arcticum*, have killed or threatened the health of livestock in some areas adjacent to the Saskatchewan and Athabasca rivers in Central Saskatchewan and Alberta since the earliest days of agricultural settlement. *Simulium arcticum* differs in many ways from the more than 100 species of black flies distributed across Canada - one of these differences being that it attacks animals rather than man. Based on work carried out in 1949-51, DDT had been effective as a larvicide for the control of black flies in the Saskatchewan River. This early work had also established DDT distribution downstream, the effect on non-target organisms and the adsorption on suspended solids. (Fredeen et al, 1953a, b) From 1968-72 field tests were carried out to find a material for black fly control to replace the persistent and, therefore, environmentally undesirable DDT. Methoxychlor looked to be the most promising since, although a chlorinated hydrocarbon, it is readily metabolized by higher animals into polar, water-soluble compounds. (Weikle et al, 1957 and Kapoor et al, 1970) Experiments were, therefore, initiated to determine the fate and effect of methoxychlor in the Saskatchewan River. The results obtained may be briefly summarized as follows. (Fredeen, 1974, 1975; Fredeen et al, 1975)

(1) Prior to methoxychlor injection in May 1972 into the system, no detectable residues of methoxychlor were found in water, sediment or biota samples despite a previous history of experimental use of methoxychlor as a black fly larvicide dating from 1968.

(2) A single 15-minute injection of 0.3 ppm methoxychlor into the North Saskatchewan River in May 1972 contaminated water and biota for a short period of time as far as 22 km (14 miles) downstream. Eight to nine days after treatment, residues of 0.05-0.10 ppm were found in sediment (sand) 21-22 km downstream.

(3) The leading edge of the treated water, indicated by buoys and the odour of the concentrate, took two hours to travel 6.5 km (4 miles) downstream. The methoxychlor levels in the water were reduced from the initial value of 0.3 ppm at the injection site by 50% (0.15 ppm).

(4) Like DDT, methoxychlor was readily adsorbed onto suspended solids (Fredeen, et al, 1975). Thus, two water samples collected 15 and 30 minutes after arrival of the leading edge of the methoxychlor 6.5 km downstream contained approximately 85 and 125 ppm suspended solids respectively. These contained 892 and 437 ppm methoxychlor respectively and the filtrates both contained 0.095 ppm. Thus, the suspended solids contained about 47% and 40% of the total methoxychlor in the water samples.

(5) Representative fauna were analyzed for methoxychlor residues but, in general, residues were non-detectable except for some larvae of aquatic insect that were disabled during actual passage of the methoxychlor "slug". Of the fish examined only the goldeye had detectable levels of methoxychlor after 8-9 days but after 17 weeks, were non-detectable. The results are summarized in Table 10.

These results, plus the general lack of reported residues in non-target organisms, sediments and water samples, indicated that methoxychlor was being rapidly degraded and, thus, removed from the environment. In view of the great concern over pesticide contamination of aquatic ecosystems, however, further experiments were carried out on the Athabasca River. This was a cooperative project between Alberta and Canada Departments of Agriculture and the Environment in 1973 and 1974. The results are of a preliminary nature since samples are still being analyzed and complete evaluation of the data must await these results. (Charnetski, 1976) The design of the experiment taken should allow a calculation to be made of the rate of loss of methoxychlor from the injected water. This was not possible with the Saskatchewan River experiment.

TABLE 10.- Methoxychlor Residues Found in Saskatchewan River Study (Fredeen et al, 1975)

Substrate	Sampling location	Sampling time	Methoxychlor content (ppm)
Water (total)	injection point	0	0.30
Water (total)	6.5 km (4 miles)	2.25 hr	0.16
Water (total)	" "	2.50 hr	0.14
Water (filtrate)	" "	2.25 hr	0.09
Water (filtrate)	" "	2.50 hr	0.08
Suspended solids	" "	2.25 hr	892.00
Suspended solids	" "	2.50 hr	437.00
Disabled insect larvae	" "	2.25-2.50 hr	17.50
Riverbed sand/silt	21-22 km (13 miles)	9 days	0.10; 0.05
Mussels	" "	8 days	ND
Insect larvae	" "	8-9 days	ND
Fish (sucker, pickerel, N. pike, sauger)	" "	8-9 days	ND
Goldeye (66% of samples)	" "	8-9 days	0.02-1.50
Goldeye	" "	17 weeks	ND

Prior to the start of the Athabasca programme, background levels of insecticide residues in fish, invertebrates, water and mud were obtained. Methoxychlor residues proved to be conspicuous by their absence. Residues were detected, however, for hexachlorobenzene, α- and γ-BHC, p,p'-DDT, DDE, and TDE, and PCBs, all at the ppb (10^{-9}g) levels.

In June 1974 the river was treated at Athabasca with 173 gallons of methoxychlor (25% EC) equivalent to approximately 0.3 mg/l. Its movement and persistence was monitored through residue analysis of water samples, fish and sediment taken at various times and space intervals within a 160 km (100 miles) downstream area and at 400 km (250 miles) with a view to determining its ultimate fate in the river ecosystem.

Movement of Methoxychlor in Surface Water

Analyses were carried out on water samples taken immediately below the surface at three locations (left and right bank and centre stream) at each of eight downstream sites situated from 0.15 km to 28.2 km from the application site. Difficulties were encountered in obtaining all appropriate samples at the critical time periods. The preliminary results indicated that uniform dispersion did not take place due to poor mixing of the methoxychlor slug and/or the flow characteristics of the river. After approximately 5.5 hours methoxychlor levels varied from 460 ppb (left bank), 161 ppb (centre stream) to 212 ppb (right bank) at a point 0.34 km downstream. The mean concentrations of methoxychlor peaked at 107, 56, and 42 ppb at 8.75, 19.5 and 28.2 km downstream. Passage of the methoxychlor slug through each site took 2.7, 4.5 and 6.0 hours respectively. Samples collected 400 km downstream peaked at 0.4 ppb and the methoxychlor took more than 76 hours to pass the sampling site.

Movement in Subsurface Water

Samples were collected approximately 20 cm above the riverbed 3.7 and 11.0 metres from shore at four downstream sites. The residue values of surface and subsurface samples were interesting insofar as they reflected the river geometry and flow characteristics. Thus, at 19.5 km from the point of application, the first methoxychlor was identified at the surface nearly two hours before the first identification at 20 cm above the riverbed 21 km downstream on the same side of the river. The fairly rapid drop of concentration encountered reflected the dilution and loss of methoxychlor to the sediment.

Methoxychlor Residues in Riverbed-load

The average methoxychlor levels found are given in Table 11. Despite the tremendous variability in values found, the relatively rapid movement downstream was obvious. Two days after the methoxychlor application, 678 ppb of methoxychlor were detected 160 km (100 miles) downstream. Within two days this value had fallen to 50 ppb and to zero at the end of 56 days. Unfortunately sampling further downstream had not been incorporated into the experimental design. The 1975 experiment extended the sampling area to 580 km (360 miles) downstream in an attempt to determine the fate of methoxychlor in the sediment.

Residues in Fish

Levels of methoxychlor in post-treatment fish (pike, walleye and longnose sucker) indicated no correlation between catch location within the first 74 km or between sexes. The levels detected would appear to be negligible in terms of risk and human consumption.

The methoxychlor treatment for black fly control was repeated in June 1975 using 64 gallons of 21% (EC) methoxychlor. This is approximately 1/3 of the dose used in the 1974 experiment. The sampling area was extended to 580 km (360 miles) in an attempt to follow movement of the bed-load material. Preliminary results indicate much lower levels than were found in the 1974 experiment even allowing for the lower dosage rate. The same general tendency for increasing concentrations downstream with increasing time after treatment was evident. This was followed by the expected general reduction in levels. One complication which will make comparison of the two years- experiment difficult is the fact that the river discharge was 16,850 cfs in 1975 as against a flow of 26,800 cfs in 1974.

Conclusions

The environmental fate of methoxychlor in an aqueous system seems reasonably well tabulated and would appear to be of little concern with regard to persistence and potential biomagnification, although the picture

TABLE 11.- Preliminary Data of Average Methoxychlor Levels in Athabasca Riverbed-load, 1974 (Charnetski, 1976)

Downstream site location from injection (km)	Methoxychlor levels (ppm, dryweight basis) Days from injection				
	0	1	2	3	4
5	5.05	4.85	0.93		0.01
20	0.03	0.34	0.04	0.01	0.02
40	0.04	0.11	0.03	0.02	0.02
60			0.02	0.02	0.02
75		0.25	0.08		0.04
100		0.13	0.10	0.05	0.03
120		0.21	0.59	0.03	0.04
140			0.58	0.06	0.03
160			0.68	0.01	0.05

at the lower end of the food chain is far from clear. Thus, there is little, if any, metabolism of methoxychlor by phytoplankton. Of the potential sinks, e.g. sediment, water, phytoplankton, fish, arthropods, etc. in the Great Lakes, residues are highest in the phytoplankton and indications are that metabolism proceeds very slowly, if at all (CCIW, 1973). Butler (1971) reported a reduction in productivity of approximately 80% at a 1 mg/l methoxychlor level. The species, incubation period and amounts of phytoplankton were, however, not reported. Nevertheless, some studies are required to determine if there is an effect on phytoplankton by methoxychlor at levels likely to be found in the environment.

The other area which needs further work is on the fate of methoxychlor in sediments. All indications are that it does not persist but the possibility of transfer downstream with subsequent dilution and build-up in a lake is a remote possibility. The present study on the Athabasca will, hopefully, resolve this problem.

NOTE: The work described under the Agricultural Watershed Study was carried out as part of the efforts of the Pollution from Land Use Activities Reference Group, an organization of the International Joint Commission, established under the Canada-U.S. Great Lakes Water Quality Agreement of 1972. Funding was provided through the Ontario Ministry of Agriculture and Food and Agriculture Canada. Findings and conclusions are those of the author and do not necessarily reflect the views of the reference group or its recommendations to the commission.

References

Butler, P. A. 1971. Proc. Royal Soc. London B. 177: 321.

CCIW. 1973. Canada Centre for Inland Waters Pesticide Survey in Lakes Erie and Ontario, October.

Charnetski, W. A. 1976. Interim Report to Alberta Black Fly Coordinating Committee

Frank, R., M. Holdrinet, H. E. Braun, G. J. Sirois and B. D. Ripley. 1976. 59th Chemical Institute of Canada Conference, June

Fredeen, F. J. H., A. P. Arnason and B. Berck. 1953a. Nature 171: 700

Fredeen, F. J. H., A. P. Arnason, B. Berck and J. G. Rempel. 1953b. Can. J. Agric. Sci. 33: 379.

Fredeen, F. J. H. 1974. Can. Ent. 106: 285.

Fredeen, F. J. H. 1975. Can. Ent. 107: 807.

Fredeen, F. J. H., J. G. Saha and M. H. Balba. 1975. Pest. Monit. J. 8: 241.

Harris, C. R. and J. R. W. Miles. 1975. Residue Reviews 57: 27

Harris, C. R. and J. R. W. Miles. 1976. Personal communication

Kapoor, I., R. L. Metcalf, R. F. Nystrom and G. K. Sangha. 1970. J. Agric. Food Chem. 18: 1145

NRC 1973. Chlordane: Its Effects on Canadian Ecosystems and its Chemistry. National Research Council of Canada No. 14094.

Roller, N.F. 1973. Survey of Pesticide Use in Ontario (Economics Branch, Ontario Ministry of Agriculture and Food)

Sanderson, M. and R. Osborne. 1976. Spring Report. International Joint Commission Office, Windsor.

Weikle, J.H. 1957. Arch. Int. Pharmacodyn. Ther. 110: 423.

Section II

DYNAMICS OF PESTICIDES IN AQUATIC ENVIRONMENTS

R. L. Metcalf, *Chairman*

ABSORPTION, ACCUMULATION, AND ELIMINATION OF PESTICIDES BY AQUATIC ORGANISMS

Fumio Matsumura

Abstract

Absorption, accumulation and elimination of pesticidal compounds were studied by using equilibrium type model ecosystems. Various factors such as the amount of pesticides, size of ecosystem, temperature, physical chemical characteristics of the pesticides, etc. were studied in relation to the rate of pesticide pick up by test organisms. The design of the model is important inasmuch as it greatly influences the outcome of pesticide accumulation studies. The pattern of bioaccumulation of pesticides is compound specific, and it appears to be possible to select certain chemicals as "benchmarks" (e.g. DDT) to arrive at relative figures for potencies for bioaccumulation. In the case of complex model ecosystems involving several biological materials, the levels of pesticides in each organism are determined by "competition" among groups of biomass for available pesticides, in addition to the "foodchain" accumulation. The rate of desorption (elimination) of pesticides appears to be species-specific. The final level of pesticides in any given species is determined by these two opposing processes: i.e. absorption and elimination reactions.

Introduction

There is no question about the phenomenon that many pesticides including most chlorinated hydrocarbon insecticides do accumulate in biological systems particularly in aquatic ecosystems. Thus it has been shown that some organisms notably the ones sitting on top of "foodchain" (particularly fish eaters) tend to accumulate high levels of pesticides. While such a generalized picture is basically correct, there are many aspects of bioaccumulation of pesticides that either clearly deviate from the general rule, or remain unexplained. For example, among fish in Great Lakes, the ones from Lake Michigan accumulate unproprotionally high levels of pesticides (Reinert, 1970) as compared to, say, the fish from Lake Erie, which is expected to contain more pesticides than does Lake Michigan. To cite another example, small fish have relatively large surface area, and also consume more food per their unit body weight than do large fish. And yet it usually is the larger fish which accumulate more pesticides in any ecosystem. While there are many other mysteries in the phenomenon of bioaccumulation, studies have just begun to understand its underlying basic mechanisms. Our attempts to study bioaccumulation started with various tests on the basic design of model ecosystems. Why model ecosystems? One may certainly ask such a question, since there are as many types of model ecosystems as the number of the scientists involved.

The answer lies first in the nature of field collected residue data: they merely represent the end results of many interactions. It is not possible, therefore, to disseminate such complex information to fathom the cause of individual reactions, nor is it wise to take the data as the clear index of bioaccumulation potency for all organisms. For instance, it is possible that the lack of bioaccumulation of pesticides beyond a certain level in any organism could rather mean the basic susceptibility of the organism to the pesticide than the low potency for bioaccumulation in that species. Second, it has become apparent that some sort of preliminary testing system for bioaccumulation potencies is necessary for many newly developed pesticides, or for those which cannot be easily studied in the environment because of the lack of proper analytical methods or because of their potential hazards (e.g. TCDD, 2,3,7,8-tetrachlorodibenzo-p-dioxin).

In designing model ecosystems we have chosen a simple equilibrium type system in which pesticides are given as dry residues on sand or absorbed on lake sediment. This method conveniently bypasses the enigma of water-solubility problems for many pesticides, and we believe, is similar to what takes place in nature. After all, the direct spraying of pestidice over open water is not a commonly employed practice today, and moreover, even when it is done in such a fashion, the pesticides in water are expected to quickly settle down to the level which does not exceed the maximum water solubility of the compound. Thus in the majority of cases, bioaccumulation takes place in the places where the sediment contains the bulk of pesticides and acts as the reservoir from which aquatic organisms gradually pick up residues.

Materials and Methods

Test organisms: Three species of fish were used. They were the northern brook silverside, *Laludesthes sicculus*, a common minnow, *Pimephales promelas* Rafinesque, and the mosquito fish, *Gambusia affinis*. The former two fish are cold water fish, while the last one is a warm water species. Three invertebrate species chosen were: an *Ostracoda* species, *Heterocypris incongruens* (adults) collected in Lake Wingra, Wisconsin, the brine shrimp, *Artemia salina* (adults) and the yellow fever mosquito, *Aedes aegypti* (4th instar larvae). The algae used were from a culture of *Anacystis nidulans* TX-20 as maintained by the method described by Batterton et al. (1972).

Pesticides: Two types of radiolabeled DDT were used. ^{14}C-DDT (ring labeled) from Amersham/Searle (40 mCi/m mole) and ^{3}H-DDT (ring labeled) from New England Nuclear (51.8 mCi/m mole), labeled TCDD (ring labeled) was provided by Dow Chemicals. It was diluted with cold TCDD to 6.1 mCi/m mole. ^{14}C-Mexacarbate (=Zectran®) was also labeled by Dow Chemicals at the benzene ring, and ^{14}C-gamma-BHC (ring labeled) was purchased from Amersham/Searle. Their specific activities were 4 mCi/m mole and 45 mCi/m mole, respectively.

Basic model ecosystem: The pesticides in ethanol or acetone were deposited in 1 g of clean sand, and the solvent was evaporated to form a dry thin-film on the surface of the sand particles. The sand was added to a jar (approximately 500 ml capacity), containing 200 ml of distilled water. The test organisms were directly transferred to the system (in the case of algae, they were concentrated by millipore filtration and resuspension and 1 ml-portion of the concentrate with known quantities in the culture medium was added to the system). The system was allowed to equilibrate for 1 to 30 days (usually 1 or 4 days), and at the end of the run the test organisms, water and sediment were assessed for radioactivities.

In desorption (elimination) studies two different approaches were tried. In one set of experiments, the sand, still containing the bulk of pesticides at the end of the uptake period (4 days), was removed and pesticides were allowed to redistribute for 4 to 8 days in the system. In another set of experiments, the test organisms were taken out at the end of uptake period (1 day), and directly transferred to another jar containing pesticide-free distilled water. Lake sediment was collected from the University Bay area, Lake Mendota, Wisconsin (Ward, 1976, type I sediment). It was air dried and the pesticide was added with solvent to 1 g of dry sediment. After the solvent was evaporated it was transferred to the system in place of the sand as the vehicle for pesticides.

Usually, the bioaccumulation and desorption experiments were conducted at room temperature (24°C) except for the ones involving cold water fish; they were conducted in a cold room which was maintained at approximately 8 to 10°C.

For radioassay materials were extracted with 4 to 1 mixture of chloroform and methanol by homogenization in the solvent mixture twice. The extracts were filtered through glass wool, the tissue debris washed with the same solvent mixture, and all washes were combined. After evaporation of the solvents the residues were directly picked up with 10 ml of liquid scintillation phosphor. In all cases, particularly the algal extracts, it was necessary to correct for quenching by using an external standardization method. In the case of some fish the above values were reexamined by means of carbonization with a Model 300 Packard Tri-Carb oxidizer and the amount of $^{14}CO_2$ was measured.

Results

Size of model ecosystems

The first obvious question on the validity of any laboratory model ecosystem is the size difference from the actual environmental ecosystem. In the first experiment, summarized in Fig. 1, we have varied the total volume of the model ecosystem. At the same time we also varied the amounts of DDT and sand proportionally. The only thing which was not changed was the size of the fish. It can be seen here that the scale of the model ecosystem does not drastically influence the rate of DDT accumulation as long as the total proportion is not changed (at least within the range tested here). On the other hand, the size of fish was found to have a strong influence. In the experiment shown in Fig. 2, the size of the ecosystem including the amount of DDT was varied according to the size of fish. Here, the level (in ppm) of DDT in the fish decreased as the size of the fish increased. It can be calculated, however, that the total amount of DDT per fish, regardless of its size was found to be roughly constant, indicating that in such a model ecosystem it is the amount of DDT picked up by the fish that is constant. A similar effect of fish size was demonstrated in a parallel experiment (Fig. 3), in which the size of the model ecosystem was kept constant.

In carrying out the above experiments, care was taken so as not to change the overall shape of the container, but in the following experiment, the depth of water was deliberately altered from the standard container while keeping the total volume of water and pesticide constant. The result (Fig. 4) shows that in this type of model the depth should not be excessively shallow in which case the fish would probably either touch or stir up the sand to cause direct contact (vs. water mediated transfer) absorption of DDT. This effect of depth may be also relevant in the case where the total volume of water was varied (Fig. 5), since these two sets of experiments produced a very similar result.

Factors influencing the degree of pickup

It has been known for some time that the higher the total amount of pesticide present in any given ecosystem, the higher the levels of its bioaccumulation. In our model ecosystem the relationship was also shown to exist (Fig. 6). The rate of increase is not, however, linearly proportional to the increase in the amount of DDT. It is noted that the relationship can be expressed by plotting log. amount of DDT applied vs. log. pickup by the organism. It is possible that it follows the Freundlich's equation, $\log Y = n \log X$

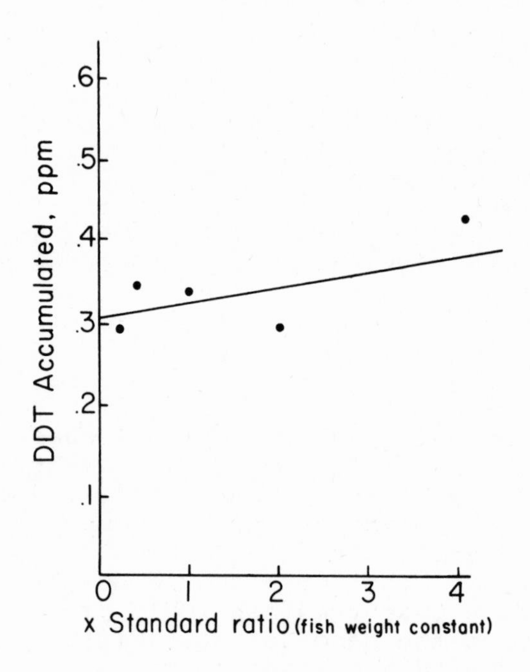

Figure 1. Effect of varying the scale of the system on the level of bioaccumulation in fish. The volume of water, amounts of DDT and sand were proportionally varied, except that the size of fish (Pimephales promelas) was kept constant. The data are expressed in multiples of the standard size model in abscissa.

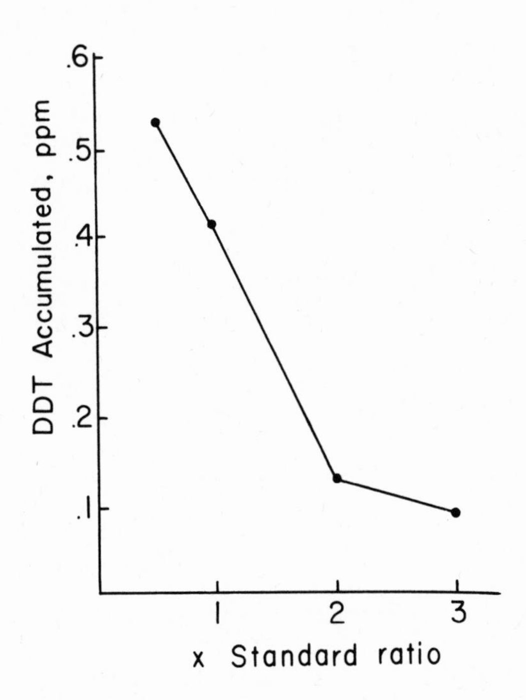

Figure 2. Effect of varying the scale of the system including the size of fish (P. promelas) on the level of DDT accumulation. The experimental conditions are identical to Fig. 1 except the variation in fish sizes.

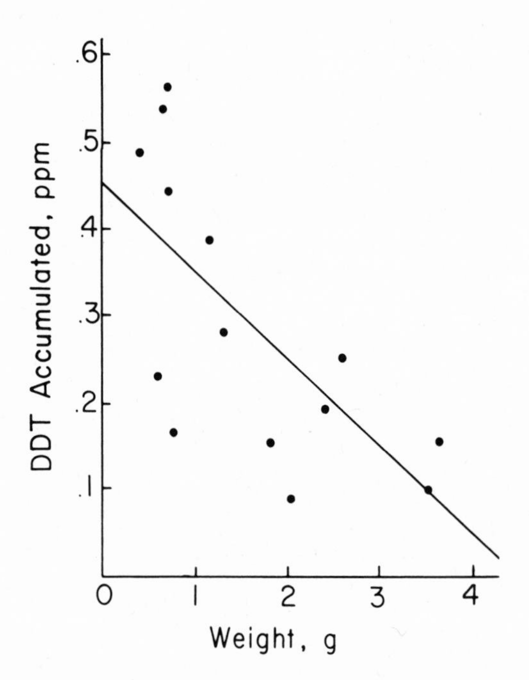

Figure 3. Effect of variation of fish sizes on the level of bio-accumulation of DDT in fish (*P. promelas*) under the standard model condition.

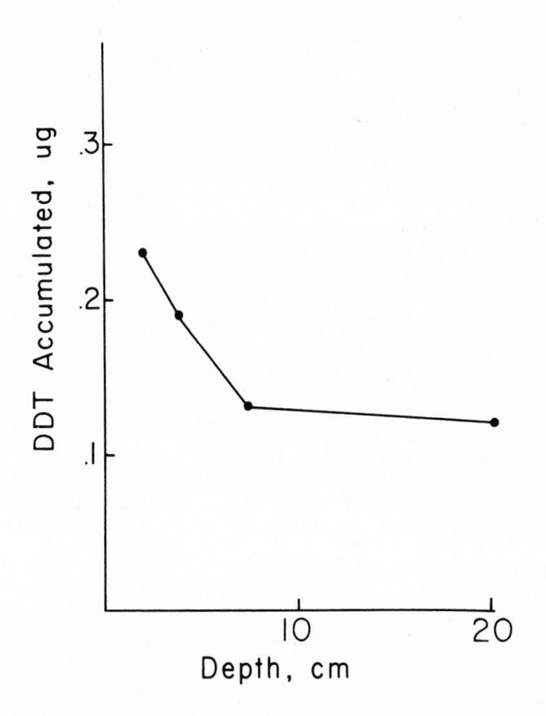

Figure 4. Effect of varying the depth of the system, at a constant volume of water on the level of bioaccumulation of DDT in *P. promelas*.

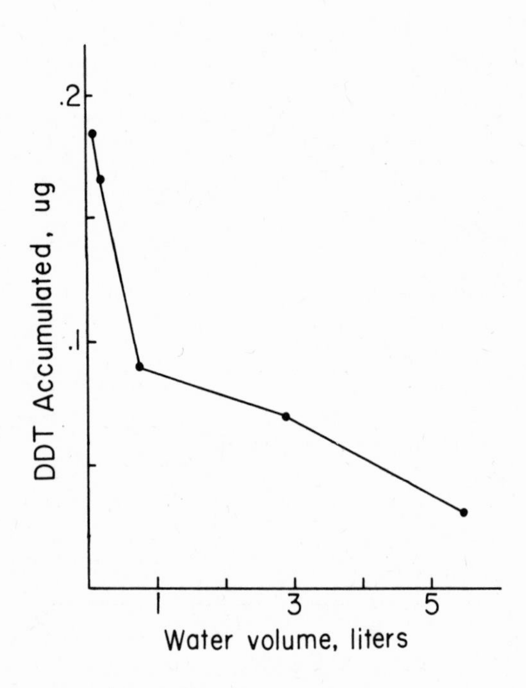

Figure 5. Variation of the water volumes affecting the level of bioaccumulation. In this experiment, unlike the ones shown in Fig. 1 or in Fig. 2, the amount of DDT and sand, and the size of fish were not changed proportionally.

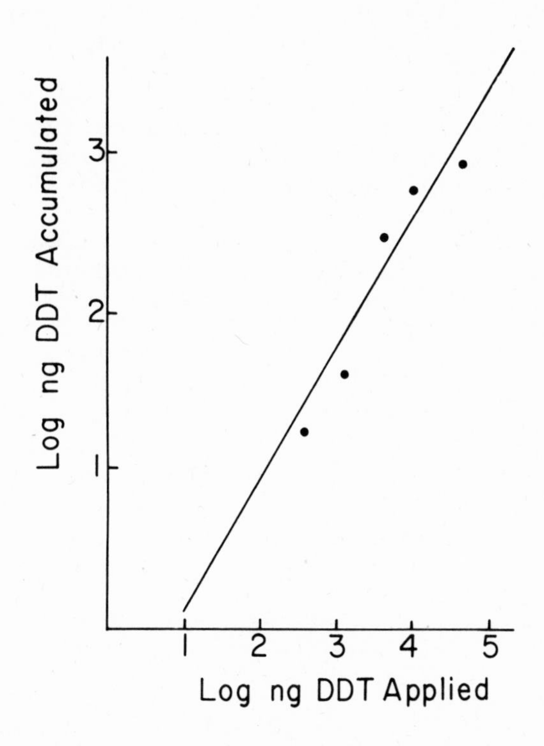

Figure 6. Effect of the changes in the amount of applied DDT on the level of bioaccumulation of DDT. (The amount of sand was also proportionally changed.)

+ log K (where n and K are constants, X is DDT applied, and Y is pickup) as suggested by Voerman and Tammes (1969).

Another important parameter in affecting the rate of pesticide pickup is temperature. It is expected that the degree of pickup is greater at higher temperatures than that at low temperatures for compounds like DDT of which water solubility increased as temperature goes up. The data shown in Fig. 7 indicates that the rate of DDT pickup increases almost linearly as the temperature rises.

Another crucial question on the relevancy of model ecosystems to the actual natural phenomenon is related to the time of pesticide exposure. Naturally any laboratory experiment cannot really extend to cover the great length of time such as years and decades. The length of laboratory tests on model ecosystems is primarily limited by the longevity of the test organisms. Even in the case where continuous multigeneration culturing is possible, usually they are limited to small organisms which do not constitute a stable member of the ecosystem.

In this type of model ecosystem, data up to a few months have been reported. In the experiment shown in Fig. 8 an attempt was made to study the long term accumulation of 4 different pesticides by ostracods. It is clear from the data that in this system an equilibrium is reached at a rather early stage (i.e. 1 to 3 days) and that the pesticide levels in ostracods did not appreciably change thereafter.

Comparison among different pesticides: Concept of "benchmark" chemicals

In the above experiment it was also shown that the degree of accumulation varies widely from one compound to another. This agrees well with our observation in nature that problems of pesticide pollution are really compound specific. That is, the pattern of bioaccumulation is largely determined by the chemical nature of the pesticide involved. The important parameters are, for example, liposolubility, partition coefficients, water solubility, stability against degradative action by biological systems and sunlight, adsorption properties to soil particles, etc.

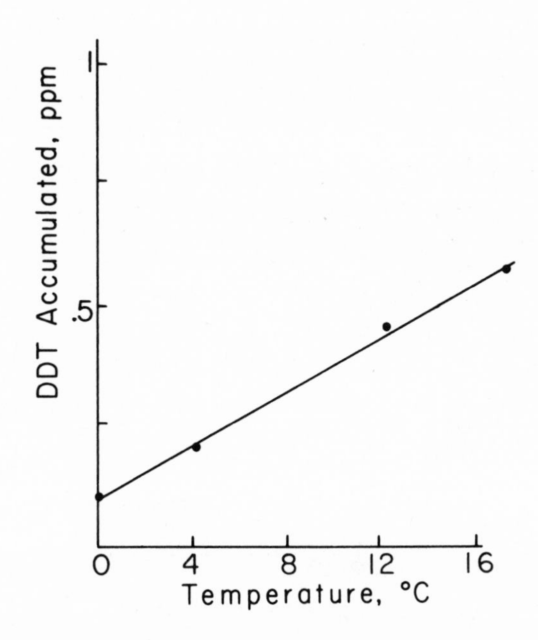

Figure 7. Effect of changes in ambient temperature on the level of DDT accumulation in P. promelas.

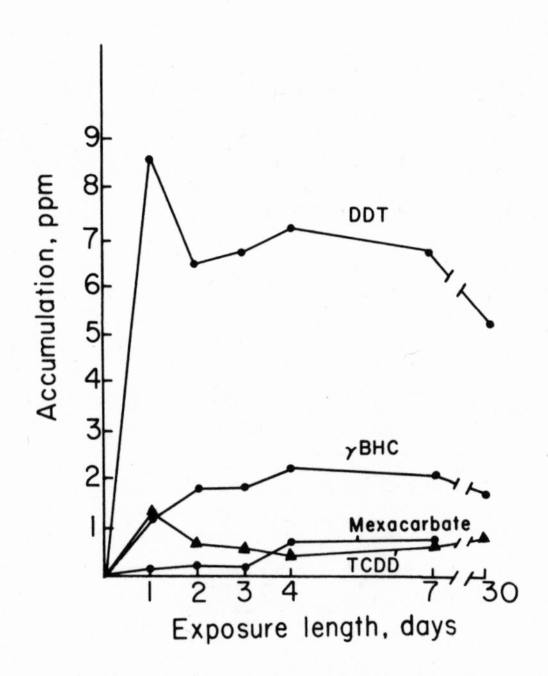

Figure 8. Long term exposure of ostracods to four pesticidal chemicals under the standard model conditions.

The most important function of model ecosystems is to serve as a prediction model for compounds which cannot be tested in the environment, or those which have been used in the field but are difficult to detect its residues in nature because of the lack of adequate analytical methods. TCDD, 2,3,7,8-tetrachlorodienzo-p-dioxin is the case in point. This compound was found among the impurities in 2,4,5-T preparation and is known to be extremely toxic, teratogenic and mutagenic. Moreover, it is a very stable compound against attacks by biological systems (Matsumura and Benezet, 1973, Ward, 1976). Despite the efforts by many scientists its residue detection has been hampered by the lack of an adequate analytical technique. Another compelling reason for the necessity to study TCDD accumulation by model ecosystems is that in this compound the most crucial question is bio-accumulation.

We now know by experience that the chemicals that cause environmental toxicological problems are the ones which are extremely persistent in nature, biologically active (toxic, carcinogenic, etc.) and are easily concentrated in biological systems. Compounds which lack any of the above qualifications usually do not play any significant role in pollution no matter how acutely toxic they are. In the case of TCDD, the question of bioactivity is indisputable. It is also stable and expected to stay in the environment for a long period of time. Thus the central question of environmental hazards must be studied from the viewpoint of "bioaccumulation".

In this study we have made attepmts to study the degree of bioaccumulation of TCDD in relation to well established chemicals such as DDT and gamma-BHC. These pesticides have been around for some time and because of their past uses and the ease of their residue detection, colossal amounts of bioaccumulation data are available. We have selected these compounds as "benchmark" chemicals against which the environmentally unknown chemical such as TCDD may be compared.

In the first series of experiments, three different species of organisms were tested separately in the standard model system (Table 1). Two important features have become apparent as the result: first, the benchmark chemicals, DDT and gamma-BHC did accumulate

Table 1. Bioconcentration of pesticides by aquatic organisms (pesticides introduced into system in the form of residues on sand)

Test Organism	Pesticide	Amount of pesticide, ug	Concentration found, ppb		Concentration factor
			Water (including food)	Test organisms	
Brine shrimp	TCDD	1.62	0.1	157	1,570
	DDT	1.79	0.5	3,092	6,184
	γ-BHC	1.47	5.2	495	95
	mexa-carbate	1.11	5.0	89	18
Mosquito larvae	TCDD	1.62	0.45	4,150	9,222
		3.24	2.40	12,000	5,000
	DDT	1.79	0.85	14,250	16,765
		3.58	1.40	30,200	21,571
	γ-BHC	1.47	6.6	1,450	220
		2.94	13.1	2,900	221
	mexa-carbate	1.11	5.45	0	0
		2.22	10.8	89	8
Fish (silverside)	TCDD	1.62	0	2	-
	DDT	1.79	2.1	458	218
	γ-BHC	1.47	1.8	2,904	1,613
	mexa-carbate	1.11	4.7	213	45

at relatively high levels in all test organisms, as compared to TCDD or mexacarbate, and second, among the test organisms the mosquito larvae showed the highest tendency to accumulate all of these chemicals. Indeed the tendency of the mosquito larvae to accumulate TCDD is really out of proportion as compared to the other two organisms. Upon a close observation on their behavior, it was found that the larvae were intensely feeding on the surface deposit of the individual sand particles. Thus the mode of pickup of TCDD and other pesticidal deposits in this particular species is not mediated via water (i.e. dissolved pesticides) but via direct contact.

Thus, this experiment was successful in demonstrating that (1) species specific factors play a very important role, and (2) chemical characteristics of pesticidal compounds do dictate the outcome of bioaccumulation with one species of test organisms. The physicochemical characteristics of these four compounds are listed in Table 2. While the most important parameter appears to be the partition coefficient in aqueous systems, both water solubility and liposolubility (as exemplified by benzene solubility as shown here) can become the rate-limiting factor to influence the final outcome. In the case of TCDD its water-solubility is only 0.2 ppb, making this compound less available to fish and brine shrimp. In addition, its poor lipid solubility prevents TCDD from achieving high levels of bioaccumulation.

It is not certain at this stage whether many pesticidal compounds have any mutual interaction in terms of the levels of bioaccumulation. We have tested this possibility in only one case involving DDT and PCB. In the experiment shown in Fig. 9 we have added PCB (2,4,5-trichlorobiphenyl) at various levels up to a few thousand times that of DDT in the standard test on DDT accumulation in fish. It was found that there is a weak antagonistic effect of PCB on DDT accumulation in the test fish.

Competition among organisms and sediments, approach to more complex systems

Admittedly the above basic system involving one organism and pesticides coated on sand is a simplistic approach, though for a start such a simple system has been necessary. As the first step toward testing more

Table 2. Physicochemical characteristics of TCDD in comparison with other insecticides

	Water solubility	Solvent solubility / Water solubility	Partition coefficient (vs. hexane)	Benzene solubility, g/100 g
TCDD	0.2 ppb	10^6	1,000[a]	0.047
DDT	1.2 ppb	10^{10}	100,000	80
mexacarbate	> 100 ppm	10^4	100[a]	-
γ-BHC	10 ppm	10^5	1,700	80

[a]Estimates.

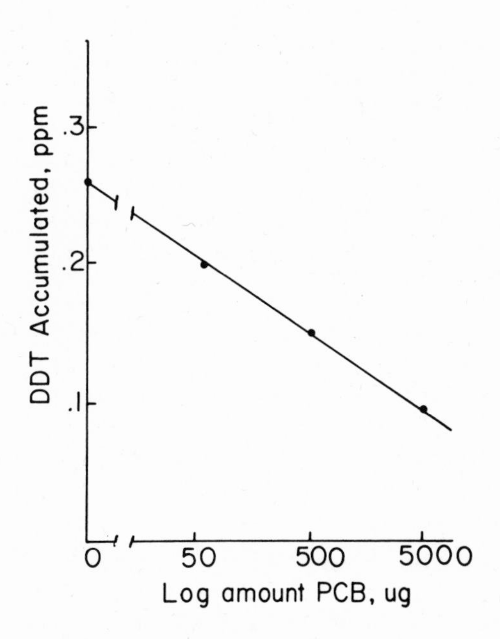

Figure 9. Effect of addition of PCBs in the same model ecosystem on the rate of DDT accumulation by the fish (*P. promelas*).

complex systems, we introduced lake sediment to the standard system. Table 3 summarizes the experimental results on ostracods exposed to 4 different pesticidal chemicals. The presence of the lake sediment caused a great change with TCDD and DDT, while in the case of gamma-BHC and mexacarbate, the change was not really significant. It must be noted here that the former chemicals have very low water solubility, and the latter chemicals have relatively high water solubilities for pesticides. Essentially an identical result was obtained when the mosquito fish *Gambusia* was used as the test organism (Table 4). Here again the effect of lake sediment was most drastic in compounds which were least water soluble.

To study foodchain accumulation the simplest approach is to design one step model for predator-prey relationship. We have therefore made an attempt to use mosquito larvae as the food source for fish (silverside). As we have noted before, the mosquito larvae accumulate extraordinary amounts of pesticides in our standard model with sand. Twenty 4th instar mosquito larvae were first allowed to accumulate pesticides for 24 hours and then fish were added. Both fish and remaining mosquito larvae (on average the fish consumed 2 to 10 larvae) were assayed for bioaccumulation after an additional 24 hours (Table 5). As expected the level of bioaccumulation in fish increased in the case of TCDD (the data are directly comparable to the ones on fish in Table 1), but in other cases, the levels of bioaccumulation actually decreased. This could be attributable to the competitive pickup of available pesticides by mosquito larvae as shown by the decrease in the pesticide levels in water (except for TCDD).

This tendency of competition for available pesticide is more pronounced when biological material with high rate of pickup are added to the system. For instance, the addition of 100 mg of unicellular algae (*Anacystis nidulans*) to the system reduces the level of bioaccumulation in fish. In this case the large surface area of algae must play a significant role in competitively picking up the pesticides in such an enclosed system.

Desorption of pesticide residues from organisms to water

In an enclosed system when the test organisms are returned to water which contain low or no level of

Table 3. Influence of the presence of lake sediment in the standard system on the level of accumulation of pesticidal chemicals in 24 hours

Pesticides	Amount of Pesticide ug	Concentration found, ppb		Concentration factor
		Water	Ostracods	
Standard test with sand				
TCDD	0.045	0.2	1302	6510
DDT	1.79	1.74	8540	4771
gamma-BHC	1.47	6.9	1200	222
mexacarbate	1.11	5.4	290	54
Lake sediment in addition to sand				
TCDD	0.045	0.002	8.8	4400
DDT	1.79	0.036	292	8111
gamma-BHC	1.47	2.07	1465	708
mexacarbate	1.11	3.71	664	179

Table 4. Influence of the presence of lake sediment on accumulation of pesticidal chemicals in fish, *Gambusia*

Pesticide[a]	Accumulation ppb			
	Water	male fish	Water	female fish
Sand, standard system				
TCDD	0.026	179	0.0113	124
DDT	2.4	14899	1.6	6691
gamma-BHC	7.0	1361	5.8	2033
mexacarbate	5.3	468	5.1	340
Lake sediment in place of sand				
TCDD	0.0002	1	0.0007	4
DDT	0.36	345	0.027	121
gamma-BHC	1.32	444	2.18	2127
mexacarbate	4.4	772	4.1	593

[a]See Table 3 for the amount of each pesticide added to the system.

Table 5. Two-step bioconcentration of pesticide by mosquito larvae, and northern brook silverside

Pesticide	Amount of pesticide, ug	Concentration found, ppb			Concentration Factor	
		Water (including food)	Mosquito larvae	Fish	Mosquito larvae	Fish
TCDD	1.62	1.3	3,700	708	2,846	545
DDT	1.79	1.1	17,900	337	16,273	306
γ-BHC	1.47	1.8	690	1080	383	600
mexacarbate	1.11	5	0	76	0	15

pesticide residues, it is known that the levels of pesticide residues in the organisms do decline, though the actual nature of such a decline has not been made clear.

In the experiment shown in Table 6 two species of organisms were first allowed to bioaccumulate pesticides for 4 days under the standard experimental conditions. The entire aquatic content of the jar, including the test organisms were then poured into another clear jar so that only the sand containing the major source of pesticides was left behind. The test organisms were kept there for 4 and 8 days. The residues remained in the test organisms were determined, thereafter. The results show that in the case of the ostracod experiment the rate of decline of residues was rather constant among 4 compounds: after 8 days the organisms retained approximately 20 to 25% of the original residues. The data were more variable in fish, where fewer individuals (1 ♂ and 1 ♀ per test) were used. However, the general tendency is that the rate of decrease of residues was much less in fish, except for gamma-BHC. The above experiments show that here probably the differences among biological systems could be more important in deciding the rate of desorption than the differences in physicochemical characteristics among the test chemicals.

In an attempt to determine the rate of pesticide desorption from different species, DDT treated organisms in the standard system for 24 hours were returned to clean water and the rates of decline in pesticide residues were monitored (Fig. 10). It is clear from the data that there are vast species differences in the rate of desorption, mosquito larvae being most rapid in eliminating DDT. Such a species difference was also clearly demonstrated by Eberhart et al. (1970) who made a close follow-up study on the rate of DDT pickup and subsequent elimination in a Lake Erie marshland ecosystem after a DDT spraying operation. They found that DDT accumulated quickly in aquatic plants and sediment particles, but they also quickly lost it. By contrast, fish generally accumulated DDT more slowly and retained them for much longer periods.

Discussion

In the effort to construct a useful model ecosystem many scientists have come up with several different types of models in the past decade. For studies on aquatic

Table 6. Absorption and desorption behavior of 4 different pesticidal chemicals by *Ostracoda* and *Gambusia* in the standard model ecosystem

Treatment	Pesticide residues in test organisms, ppb			
	TCDD	DDT	gamma-BHC	mexacarbate
Ostracoda				
Absorption, 4 days	86	19,613	2,953	550
Desorption[a], 4 days	34	10,190	1,173	563
8 days	26	3,390	543	188
Gambusia				
Absorption, 4 days	152	10,800	1,695	405
Desorption[a], 4 days	79	13,165	850	263
8 days	76	25,270	328	688

[a] In this set of experiments the pesticide source on sand was removed after 4 days absorption period. The system was re-equilibrated for 4 and 8 days thereafter.

ecosystem, where the most significant bioaccumulation of pesticides takes place, essentially there appear to be three basic types of models. First is the "self-sustaining type" ecosystem. The most well known example is the one designed by Dr. Metcalf and his colleagues at University of Illinois, Urbana (e.g. Kapoor et al., 1970). The second type is what we call "dynamic-circulating" system in which a constant and defined level of the pesticide usually suspended or dissolved in water is delivered to the tank containing the test organisms (e.g. Chadwick and Brockson, 1970). The third type is the one I have just described herein: organisms are kept in a tank and pesticides are usually given as deposits prior to the introduction of organisms. Here the last type is designated as "equilibrium type." A similar system was also utilized by Isensee and Jones (1971) in studying TCDD accumulation in the laboratory.

In such a system pesticidal residues are largely distributed according to the partitioning behavior of the chemical in relation to the biological systems. An equilibrium is established relatively fast (in a matter of days). All the processes of redistribution thereafter are slow and quite often relatively insignificant. These latent factors are metabolic and photochemical changes on the pesticidal molecules, biological growth, including that of test organisms as well as microorganisms, adsorption and eventual elimination of pesticide from circulation, evaporation of pesticides and water, increase in organic matter in water and sediment as the result of biological activities, excess food and death of organisms, gradual changes in redox potential, pH, dissolved oxygen, electrolyte contents, etc. Thus in the case of stable pesticidal chemicals such as DDT, what one is actually measuring in such an equilibrium model is largely the results of the initial distribution of the pesticide through partitioning. In this type of model therefore, the experimental design calls for a relatively short term testing: reasonable time periods for testing would be the matter of days and weeks, but not months and years.

In the equilibrium systems two basic reactions are involved in the process of partitioning: (a) dissolving the solid pesticide deposit, either from sand or glass surface into water, and (b) partitioning the pesticides from water to the test organisms. The former process is clearly controlled by water solubility of the compound, and the latter process is determined by partition coefficient and lipsolubility (these two terms are not synonymous, see Table 2 benzene solubility for liposolubility) as attested to by the compound specific differences observed in the rate of bioaccumulation (Table 1, 2). When lake sediment or soil is substituted for sand (or glass surface) the former process is no longer limited by only water solubilities of the pesticides: i.e. the process of release of pesticides to water becomes more complex. Indeed, sediment can irreversibly absorb pesticides and thereby eliminate them from circulation or its organic matter can act as another quasi-lipid pool to which pesticides can partition against water.

At any rate, in most cases, water acts as the mediator of initial pesticide translocation in this type of model, except for bottom dwellers and scavengers (such as mosquito larvae) for which transport of pesticides is direct contact. Even with free swimmers, when the depth of the water becomes shallow the chance of direct contact and resulting high rates of pickup increases (Fig. 4).

Knowing that the amount of pesticide dissolved at any given time controls the eventual level of bioaccumulation, at least for free swimming organisms, the result of the temperature experiment (Fig. 7) becomes self explaining: i.e. at higher temperatures higher amounts of pesticides can be dissolved in water.

Despite some useful data generated by the use of model ecosystems, there is still a vast gap between the laboratory data and the field observation. The above example on temperature effects does not directly translate into high levels of pesticide accumulation in southern aquatic ecosystems as compared to northern systems. Naturally other factors such as higher rate of degradation at high temperatures, more vegetation and organic matter in aquatic systems in the South, etc. play significant roles as well. In the case of TCDD unusual rates of pickup by mosquito larvae have been cited as the evidence for bioaccumulation by some people, while the picture could be quite different in nature where TCDD is tightly bound to aquatic sediment and its availability to mosquito larvae is quite limited.

The usefulness of laboratory model experiment rather lies in its possibility for future studies on the basic information on bioaccumulation mechanisms. Only

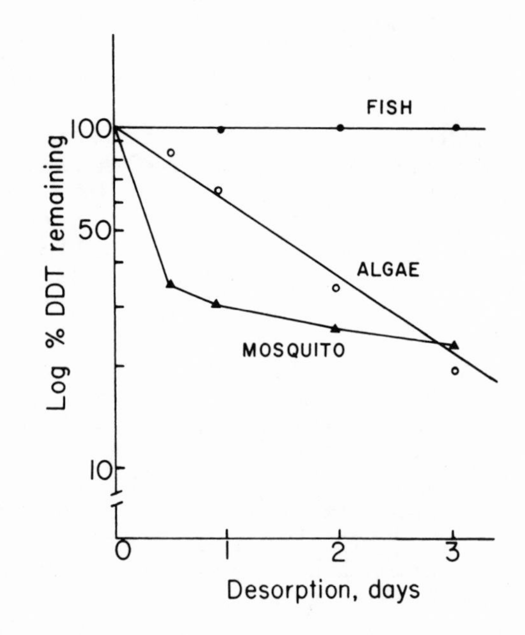

Figure 10. The rate of desorption of DDT in three aquatic organisms. They were first kept in the standard model for 1 day and then transferred to the same volume of clean distilled water for the specified desorption period.

through such laboratory experiments can each factor involved be analyzed and determined.

Acknowledgment

This work could not have been completed without the assistance of Dr. Herman J. Benezet, Mr. James Preysz and Ms. Keiko Matsuda. This work is supported by the College of Agricultural and Life Sciences, University of Wisconsin, Madison, Hatch Project 822.

References

Batterton, J. L., G. M. Boush and F. Matsumura. (1972). Science 176: 1141-3.

Chadwick, G. G. and R. W. Brockson. (1969). J. Wildlife Management 33: 693.

Eberhart, L. L., R. L. Meeks, and T. J. Peterle. (1971). Nature 230: 60.

Isensee, A. R. and G. E. Jones. (1971). J. Agr. Food Chem. 19: 1210-1214.

Kapoor, I. P., R. L. Metcalf, R. F. Nystrom, and G. K. Sangh. (1970. J. Agr. Food Chem. 18: 1145-1151.

Matsumura, F. and H. J. Benezet. (1973). Environmental Health Perspectives, National Insitute of Environmental Health Sciences, September, 1973, 253-258.

Reinert, R. E. (1970). Pesticide Monitoring Journal 3: 233.

Voerman, S. and P. M. L. Tammes. (1969). Bull. Environ. Contam. Toxicol. 4: 271-277.

Ward, C. T. (1976). M.S. Thesis, Department of Entomology, University of Wisconson. pp. 21-103.

ELIMINATION OF PESTICIDES BY AQUATIC ANIMALS

M.A.Q. Khan

ABSTRACT

The absorption and elimination of chlorinated hydrocarbon insecticides, DDT, Lindane, and cyclodienes, seem to occur simultaneously in fish. Continuous long-term exposure for several months may show higher levels of these insecticides, especially the most lipophilic DDT, in fat but the results of short-term exposures indicate that accumulation in tissues involves factors other than just the lipophilicity of the insecticide and the total fat contents of the tissue.

Transfer of the pre-exposed animals to insecticide-free water results in the elimination of the absorbed chemical. The rate of elimination depends on the water-solubility of the insecticide. Lindane, chlordane, and photodieldrin are more rapidly eliminated than more lipophilic DDT. In fish tissues of all organs eliminate at about similar rates. Renal, bronchial, biliary (fecal), and integumentary routes are all involved in the excretion of these insecticides.

INTRODUCTION

Contamination of aquatic environments with lipophilic organic chemicals such as chlorinated hydrocarbon pesticides, polychlorinated biphenyls, hydrocarbons in fuel oils, etc. can result in the absorption and accumulation of these chemicals by aquatic organisms. The dynamics of this phenomenon have been described earlier in this book (Haque et al., 1977; Matsumura, 1977). The factors so far known that influence the absorption include lipophilicity, water solubility, and partition coefficients of these chemicals. The trophic transfer of these chemicals can lead to high levels of these contaminants in organisms at the top of food chains (Metcalf, 1977; Craig and Rudd, 1974; Woodwell et al., 1967; Rudd, 1964). It has also been well-documented that if further contamination of the aquatic environment is discontinued or if the contaminated organisms such as fish are transferred to insecticide-free water, they rapidly eliminate the absorbed chemicals from their tissues (Gackstatter, 1966; Gackstatter and Wise, 1967; Grzenda et al., 1970, 1971, 1972; Tooby and Durbin 1975; Khan et al., 1975; Macek, 1970). It is this dynamic aspect of pesticides in aquatic animals that will be covered in the following pages.

1. Absorption of chlorinated hydrocarbon insecticides

The fish can absorb insecticides directly from water as well as by ingesting the contaminated food. The more lipophilic the chemical, the more readily it can be taken up by aquatic organisms especially if the concentration of this chemical exceeds its water solubility. Goldfish exposed to 10 ppb absorbed the following amounts of the insecticide (% of the dose): DDT - 74; dieldrin - 54; and lindane - 11 (Gackstatter, 1966). This was related with the liposolubility of these compounds. The size, age, and sex of the same population of the species can affect the absorption rate (see Kenaga, 1975). The same chemical can be absorbed at different rates by different species. For example, during a 24-hour exposure to 3 ppb of alpha-chlordane goldfish absorbed 56% and bluegills only 42% of the insecticide (unpublished data, this laboratory).

The gills and the surrounding soft parts of the body are the primary sites of absorption (Gackstatter 1966). For example, Atlantic salmon, *Salmo solar*, exposed to 1 ppm DDT absorbed it within 5 minutes through gills and showed up to 31 ppm DDT in liver and spleen in one hour (Premdas and Fernando, 1963). The brown trout exposed to .1 and .5 ppm DDT absorbed mainly through gills and distributed via blood to all tissues (Holden, 1962).

The mechanisms of transport of organochlorine insecticides across membranes of gills, blood capillaries, blood corpuscles, intestinal epithelia, and target tissues have not been worked out. The chemical whether ingested or absorbed via gills rapidly appears in the blood (plasma albumins and erythrocyte hemoglobins) and becomes distributed in tissues of all soft organs (Tables 1, 2, 3, 4) (Moss and Hathway, 1964). The rate of dieldrin removal from blood, which apparently serves a transitionary function, is initially rapid and then becomes approximately logarithmic.

2. Elimination of pesticides by fish

The elimination of the absorbed insecticide seems to occur simultaneously with its absorption. According to Tooby and Durbin (1975) it is the balance between these two processes that determines the fate of the chemical in tissues. Grzenda et al. (1970, 1971, 1972) have raised the question that the storage of DDT, dieldrin, and lindane in tissues is not related with their content of fat. Prolonged continuous exposure to DDT, which is more lipophilic than dieldrin, can result in high levels of DDT and/or its metabolites in visceral fat (Bridges et al., 1963; Cope, 1969). This has been observed also with dieldrin (Grzenda et al., 1971) and endrin (Mount and Putnick, 1962).

When the pre-exposed fish are transferred to clean water or fed uncontaminated food, the concentrations in tissues may decrease except for the fat where it may increase (Tables 2, 3; Fig. 1) (Schoenthal, 1963).

The rates of elimination of pesticides depend on their water solubility. About 50 to 90% of the absorbed pesticide can be eliminated within four weeks. Malathion, Simazine, and Bayer-73 are eliminated in one to three days (Macek, 1970; Statham and Lech, 1973). The elimination of organochlorines by goldfish in the decreasing order is: lindane > dieldrin > DDT and this correlates with their water solubilities. Goldfish eliminated 90% of the absorbed lindane and dieldrin in, respectively, 2 and 16 days, while no more than 50% of the absorbed DDT was eliminated in 32 days((Grzenda et al, 1971, 1972; Grackstelter and Wise, 1967; Tooby and Durbin, 1975). About 98% of the absorbed endosulfan is eliminated, as its diol, in 14 days (Schoethger, 1970; Farbwerk, 1971; Gorbauch and Knauf, 1972). The channel catfish, *Ictalurus punctatus*, eliminated 50% of the

Table 1. Concentration of dieldrin in various organs of goldfish fed 0.812 and 6.1 nanomoles of ^{14}C-dieldrin for various periods of time (Grzenda et al.,1971)

Organs/ Tissue	Pico-mole dieldrin/g wet wt. Total Exposure: days of continuous feeding and total dieldrin fed (nM)								Mean Relative Distribution for .812nM fed for 192 days*
	8 days		16 days		64 days		128 days		
	.57nM	4.25nM	1.14nM	8.5nM	4.55nM	34nM	9.09nM	68nM	
Blood	5	15	7	51	13	91	18	355	1.1±.1
Skin	14	63	14	90	39	140	25	378	2.6±.3
Nerve	45	237	34	445	46	461	56	878	5.4±.7
Muscle	4	26	6	55	9	75	14	244	1.0±.1
Gills	5	40	9	84	21	132	24	268	1.7±.1
Brain	16	80	17	235	36	291	49	725	3.4±.1
Gallbladder	19	117	28	210	45	281	48	583	3.9±.3
Spleen	8	63	18	209	20	220	32	702	2.2±.1
Liver	13	107	20	231	22	331	58	602	3.4±.2
Stomach	9	74	42	140	26	213	52	648	3.3±.5
Intestine	13	108	36	314	25	278	40	613	3.3±.3
Intestinal Contents	46	220	25	466	45	538	48	1236	4.2±.7
Kidneys	10	75	16	176	24	257	43	625	2.6±.2
Mesenteric Fat	87	-	571	310	272	2141	152	1423	21.5±12.7
Immature Ovary	10	21	10	90	38	-	-	1836	2.0±.6
Testes	-	333	54	413	122	1737	93	7209	8.1±1.9

Table 2. The concentration of insecticides in the tissues of fish at various times following their exposure to an initial concentration of 70ppb of each insecticide (Gackstatter, 1966)

Organs	insecticide concentration = ppm*								
	DDT			dieldrin			lindane		
	0 days	2 days	4 days	0 days	2 days	4 days	0 days	2 days	4 days
Liver	11.00	3.15	3.25	14.75	4.15	4.3	9.60	2.15	.80
Kidneys	7.45	1.50	1.60	4.10	1.60	1.6	3.05	.90	.40
Gills	5.80	4.50	5.55	5.00	6.70	7.2	6.40	2.75	1.05
Gallbladder	.30	17.55	14.10	1.65	-	-	2.40	1.50	2.50
Stomach	2.40	1.20	1.20	2.15	0.90	.8	1.90	.65	.35
Intestine	2.80	1.10	1.60	4.75	3.80	4.1	5.50	3.40	1.50
Pyloric ceca	5.85	13.10	10.25	9.10	4.50	7.6	5.70	2.40	1.80
Skin	11.90	12.10	9.50	1.35	1.05	1.2	1.60	.40	.20
Brain	3.75	3.50	6.05	4.15	11.10	2.8	3.60	1.80	2.40
Muscle	.95	.55	1.15	0.95	.45	.4	1.20	.25	.15
Gonads	2.65	1.35	50.60	2.20	.60	.9	2.45	.80	.35
Fat	-	58.45	-	32.50	69.00	41.0	-	-	-
Spleen	4.00	-	3.00	-	2.60	-	-	-	-
Carcass	2.20	2.45	2.60	2.45	2.50	2.9	3.10	1.60	.70

* most of the values are averages of 2 fish. Bluegills (43 to 71 g wt) were used for DDT and dieldrin, redear (33 to 39 g wt) for lindane.

Table 3. Concentration of insecticides in tissues of goldfish after 4 days following their exposure to 30ppb of each insecticide

Tissues	ppm insecticide*					
	DDT		dieldrin		lindane	
	0 days	4 days	0 days	4 days	0 days	4 days
Kidneys	4.75	3.10	3.55	0.65	6.10	0.00
Gills	2.50	1.85	2.40	0.40	0.65	0.00
Gill area	6.15	6.65	8.55	0.55	11.85	0.00
Gallbladder	1.80	6.30	2.55	1.40	2.35	1.00
Brain	5.45	5.20	6.60	0.95	2.05	0.00
Heart	3.50	2.55	2.80	0.60	1.50	0.00
Air sac	2.05	1.40	1.70	0.35	2.50	0.00
Gonads ♂	10.30	23.20	7.80	1.30	3.70	–
♀	2.60	3.60	3.40	3.80	3.60	0.10
Muscle						
Lateral tail	2.25	1.55	1.85	0.40	0.90	0.00
Anterior epaxial	2.00	1.25	1.70	0.30	0.80	0.00
Posterior lateral	2.15	1.30	1.70	0.30	0.75	0.00
Lateral-visceral	2.55	1.25	2.35	0.45	1.65	0.00

* Average of two fish (15.1 to 23.5 g wt).

Table 4. Distribution of ^{14}C-α-chlordane in tissues of goldfish exposed to 4ppb α-chlordane for 24 and 48 hours (Moore et al., 1977)

	Concentration in tissues: ppm equivalent of chlordane	
	average	
Tissue[a]	24 hr	48 hr
Gills	.064	.043
Air sac	.058	.099
Tail muscle	.019	.074
Eggs	.056	.134
Bile	.119	.089
Gallbladder	.116	.148
Liver	.022	-
Kidney	.061	.154
Stomach	.032	-
Intestine	.063	.127
Intestinal content	.055	.094
Brain	.140	.086
Spinal cord	.053	.104
Heart	.078	-
Abdominal fat	1.042	1.572
Skin	.046	.202
Blood	.008	-

[a]3 fish (85 gm/fish) exposed, in 2 L water, separately for 24 hours at room temperature

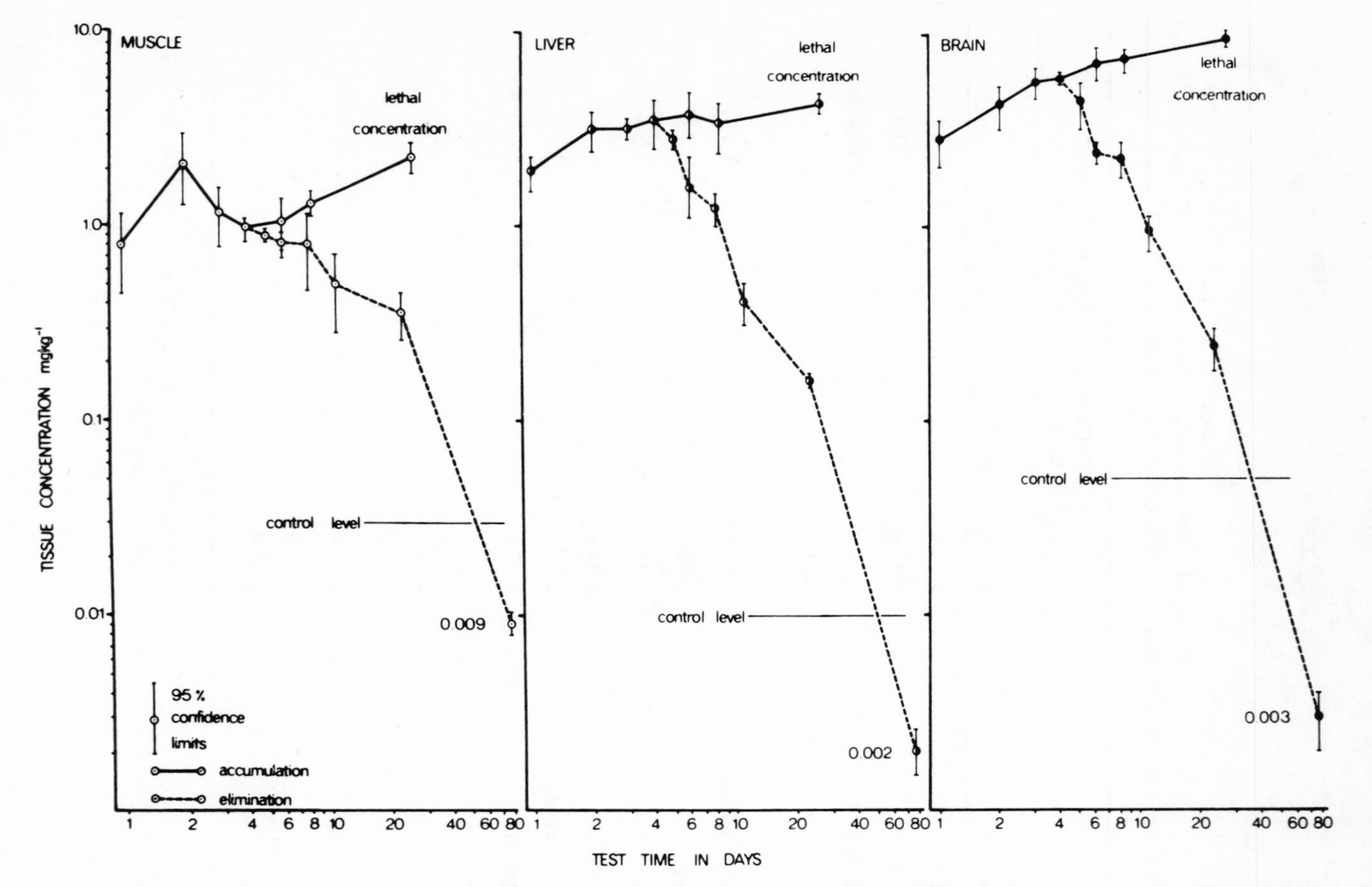

Fig. 1. Accumulation and elimination of lindane in tissues of yearlings of the rainbow trout exposed to 26 ppb lindane (Tooby and Durbin, 1975).

absorbed dieldrin in 9.6 days (Argyle et al., 1975). Bayer-73 is also rapidly eliminated by the rainbow trout, *Salmo gairdneri*, (Statham and Lech, 1973). The half-life values of various pesticides in rainbow trout are shown in Table 5. Mosquitofish, *Gambusia affinis*, eliminated 75% of the absorbed DDT (Meikel et al., 1975) and endrin in 7 days (Ferguson et al., 1966). While goldfish eliminate DDT slowly over an extended period, bluegills eliminate it rapidly within the first four recovery days and then the concentration remains constant. Bluegills, like goldfish, eliminate lindane more rapidly than dieldrin (Gackstatter, 1966). Goldfish eliminated chlordane (Table 6) much faster than photodieldrin (Fig. 2).

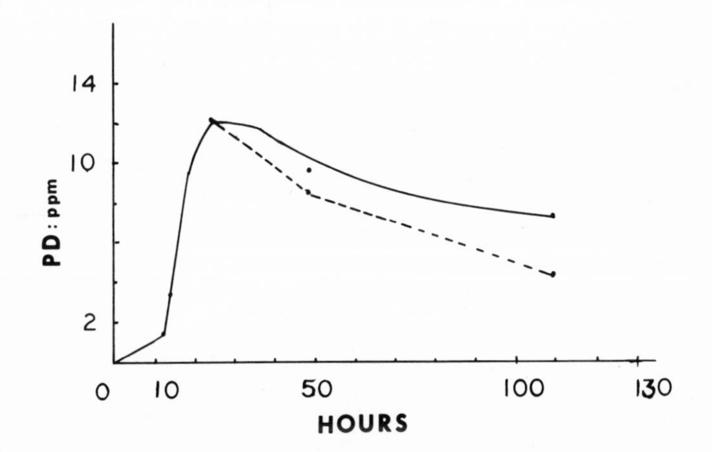

Figure 2. Absorption (during continuous exposure to 3 ppb photodieldrin, solid lines) and elimination (on transfer to clean water, broken lines) of photodieldrin by goldfish (Khan et al., 1975).

Table 5. Persistence of pesticides in rainbow trout (Macek, 1970)

Persistence	Pesticide
< 1 day	Malathion
< 2 days	Lindane
< 3 days	Simazine
< 1 week	Diazinon, Dursban, Azinphosmethyl Parathion, Methoxychlor, 2,4-D
< 2 weeks	Dichlorobenil
< 3 weeks	Diquat, Endothal
1 month	Heptachlor, Dieldrin
4 months	Sodium arsenate
> 5 months	DDT
> 6 months	DDD, Campheclor

Table 6. Absorption and elimination of chlordanes (1:1 mixture of α- and γ-isomers) by goldfish individually exposed to 3.44 ppb (4 μg; 49,800 DPM) in 750 ml of water (Moore et al., 1977)

Observation Time	Chlordanes in fish			
	μg/fish (average)	ppm		% of the initial dose/fish (average)
		range	average	
I Absorption*				
12 hr	1.18	.18-.20	.19	40.18
1 day	.91	.14-.15	.14	29.40
2 days	1.11	.13-.20	.16	34.65
3 days	1.52	.22-.23	.23	48.30
II Elimination**				
0 hr	2.00 (1.09)	.35-.37	.36 (.71)	58.88 (32.06)
1 day	1.25 (1.00)	.11-.37	.25 (.54)	45.00 (28.36)
2 days	1.23 (-)	.12-.27	.22 (-)	39.60 (-)
4 days	.90 (.72)	.11-.27	.18 (.49)	30.40 (21.17)
8 days	- (.63)	-	- (.38)	- (18.53)
11 days	- (.11)	-	- (.08)	- (3.24)

* 6.3 g/fish, exposed at 25 ± 2°C.

** 5.4 g/fish, exposed for 24 hr and then transferred to clean water (= 0 hr); values in parenthesis for smaller fish (1.7 g/fish), both these experiments carried out at 26 ± 2°C. Percent of the initial dose calculated from μg/fish.

The studies with individual organs of goldfish showed that half-life of dieldrin was: intestinal contents - 8 days, brain and liver - 27 days, intestine - 30 days, all other tissues - 19-24 days (Grzenda et al., 1972). Half-life of lindane in muscle, liver, and brain lies between 7 and 10 days (Tooby and Durbin, 1975).

Starvation of the pre-exposed fish during the recovery period results in rapid elimination of dieldrin (Grzenda et al., 1972). There may be higher levels of dieldrin in the brain, liver, and blood, as compared with the unstarved fish, with no increase in dieldrin levels in fat. The dieldrin may be remobilized from the fat and re-distributed in other tissues or excreted (Grzenda et al., 1971, 1972). The fat apparently does not provide a protection against lipophilic organochlorines.

The excretion of organochlorines can involve liver, kidneys, gills, and integument. The insecticide and its metabolite(s) pass from liver to the gallbladder and from there to duodenum. The insecticide concentrations were lower in the stomach as compared to those in the pyloric ceca and intestine showing a concomitant decrease in concentration in liver (Table 2). This indicates biliary and fecal elimination of organochlorines. The elimination of organochlorines through gill membranes into the external medium is apparently very rapid. The gills, and due to similar elimination, kidneys show low levels of insecticides (Table 2). The elimination of the lindane residues from brain is as rapid as that from liver of the rainbow trout, 95% of the accumulated residue lost in 19 days while only 65% lost from the muscle during this period followed by a rapid elimination thereafter (Tooby and Durbin, 1975).

Other foreign chemicals are also eliminated by fish. This includes elimination of hydrocarbons in fuel oils by killifish, *Fundulus similus*, and other marine fish (Lee et al., 1972); of antimycin by the bullheads, *Ictalurus nebulosus* (Schultz and Herman, 1976); of methylmercury by the rainbow trout (Giblin and Massero, 1973); and so on.

3. Elimination of pesticides by aquatic invertebrates

The fresh water flea, *Daphnia magna*, pre-exposed to dieldrin eliminated 90% of the absorbed insecticides in 3-4 days on transfer to insecticide-free water (Fig. 3). The elimination is more rapid if the exposure is in the absence of food (yeast or algae). Photodieldrin elimination by *D. pulex* is also very rapid (Fig. 4)

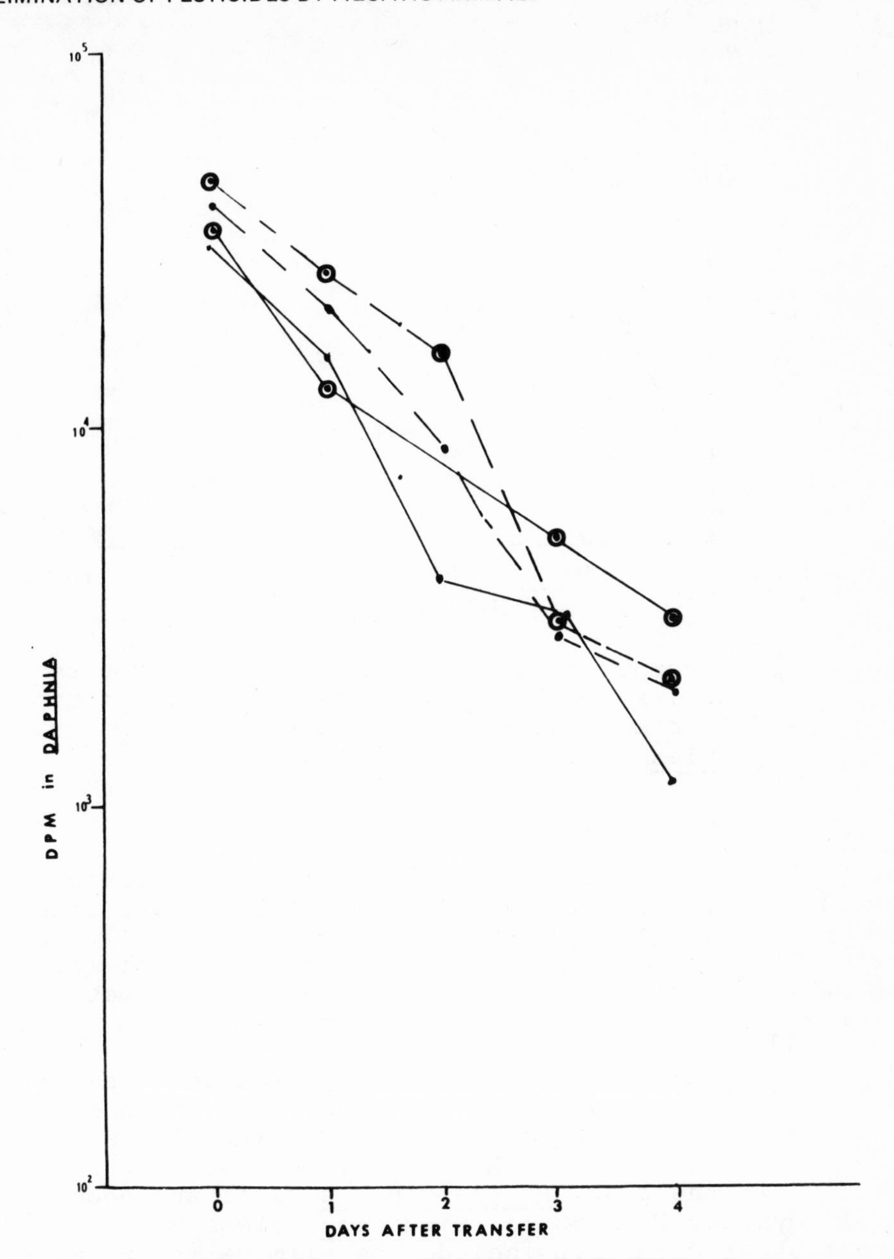

Figure 3. Elimination of dieldrin by Daphnia magna pre-exposed to 0.5 ppb dieldrin for 24 hours with algae (solid lines) or yeast (broken lines) on transfer to clean water. The values for algae and yeast were obtained, in each case, with 100 cells (•——•) and 1000 cells (o——o)/ml (Stanton et al., 1973).

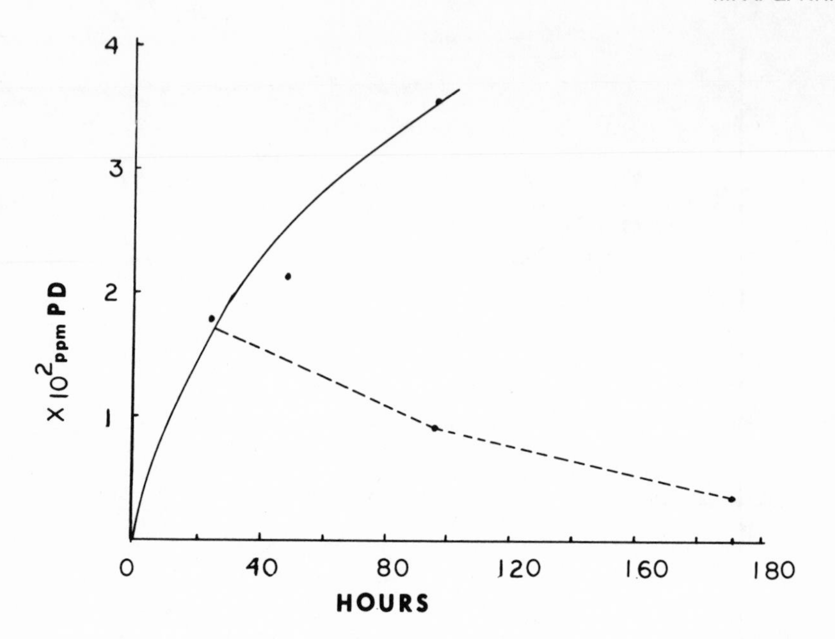

Figure 4. Absorption (during continuous exposure to 3.33 ppb, solid line) and elimination (on transfer to clean water) of photodieldrin by Daphnia pulex (Khan et al., 1975).

Chlordanes are also rapidly eliminated by the latter species (Fig. 5). DDT has been reported to be eliminated by the shrimp, Euphausia pacifica (Cox, 1971). More systematic studies on the elimination of organochlorines and other insecticides by daphnids and other aquatic invertebrates need to be carried out.

Other xenobiotics have been reported to be eliminated by aquatic invertebrates. Hydrocarbons in fuel oils are eliminated by the mussel, Mytilus eludis (Clark and Finley, 1972; Lee et al., 1972a; Anderson et al., 1974); the clam, Rangia cunacta (Neff and Anderson, 1974); oyster, Crassostrea virginica (Stegman and Teel, 1973); other chemicals include the elimination of methylmercuric chloride by oysters (Cunningham and Tripp, 1973, Smith et al., 1975).

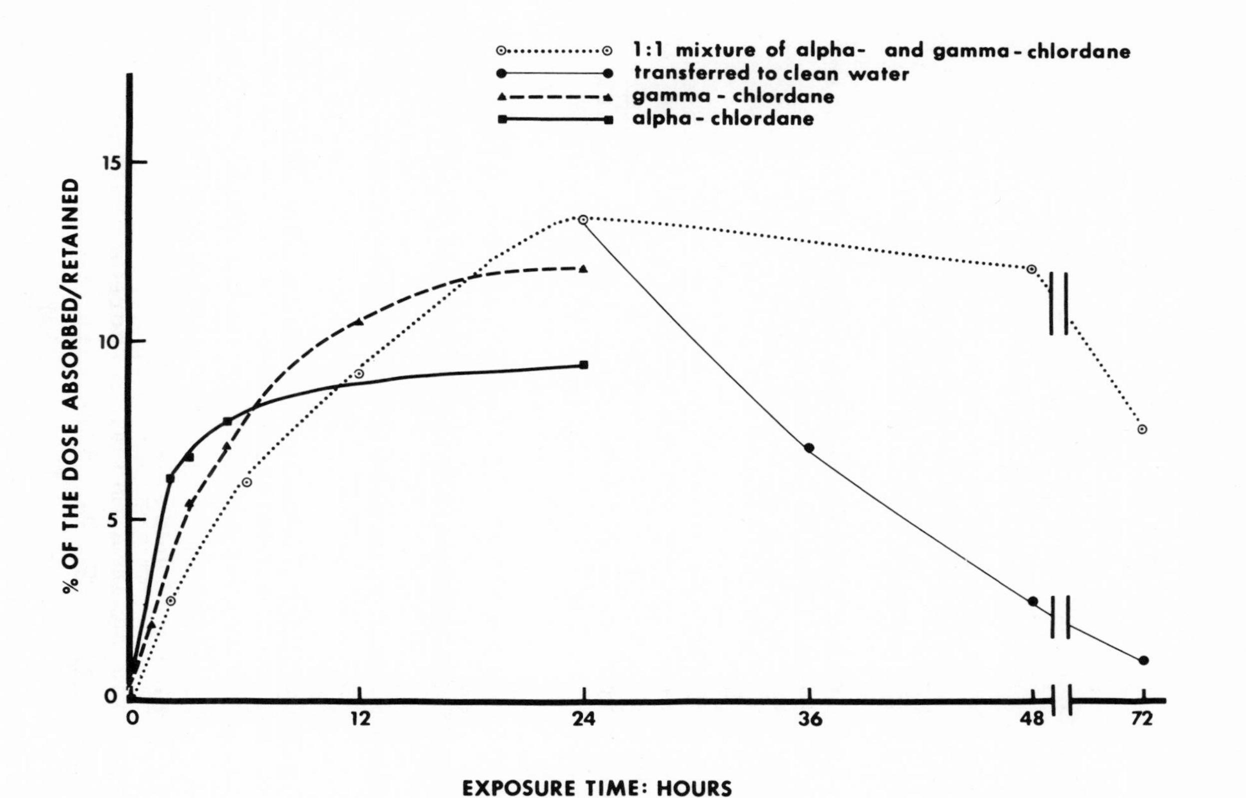

Figure 5. Rate of absorption and elimination of α - and γ-chlordanes and of a 1:1 mixture of the two isomers by Daphnia pulex (60 daphnids, 1.65 mg to 2.04 mg dry wt.) during continuous exposure to an initial concentration of 0.5 ppb (100 nanogram in 200 ml of water) and following their transfer to clean water. The percent of the initial dose absorbed is equivalent to nanograms absorbed by 60 daphnids (Moore et al., 1977).

CONCLUSIONS

The aquatic invertebrates and fish seem to be in a state of dynamic exchange of pesticidal chemicals with their environment. The simultaneous occurrence of absorption and elimination of foreign chemicals suggests that internal homeostatic mechanisms tend to maintain the balance in favor of outward movement of lipophilic xenobiotics, and all organs appear to be responsible for this. This balance or equilibrium may be temporarily upset by high pesticide concentration in the medium; but decontamination of the environment can immediately result in the elimination of the stored chemical. The storage of the lipophilic organochlorines does not depend on the fat content of the particular tissue. The elimination depends on the degree of water or lipid solubility of the chemical. More studies about the processes such as the physiological-biochemical mechanisms of the animal, physico-chemical properties of the chemical are needed to provide a clear understanding of this subject.

Aquatic animals are very sensitive to chemicals directly absorbed from water and thus cannot be treated with high concentrations of the experimental toxicant. The amounts of *in vivo* metabolic products are therefore low and make it very difficult and expensive to quantitatively analyze the extent of metabolism of insecticides. Therefore, the information provided here has not emphasized whether the elimination involves the alteration of the chemical. This subject is covered elsewhere in this book (Khan et al., 1977).

REFERENCES

Anderson, J.W., J.M. Neff and S.R. Petrocelli. 1974. In: Survival in Toxic Environments (M.A.Q. Khan and J.P. Bederka, Eds.). Acad. Press. p. 83-121.

Argyle, R.L., G.C. Williams and C.B. Daniel. 1975. *J. Fish. Res. Bd. Canada* 32: 2197.

Bridges, W.R., B.J. Kallman and A.K. Andrews. 1963. *Trans. Am. Fish. Soc.* 92: 421.

Clark, R.C. and J.S. Finley. 1975. *Fishery Bull.* 73: 508.

Cope, O.B. 1966. *J. Appl. Ecol.* 3:33 (supplement on pesticides in the environment and their effects on wildlife).

Cox, J.L. 1971. Fishery Bull. 69: 627.

Craig, R.B. and R.L. Rudd. 1974. In: Survival in Toxic Environments (M.A.Q. Khan and J.P. Bederka, Eds.). Acad. Press. p. 1-24.

Cunningham, P.A. and M.R. Tripp. 1973. Marine Biol. 20:14.

Farbwerke Hoechst, A.G. 1971. Thiodan and the environment. Technical Bulletin translated by Hoechst, U.K.

Ferguson, D.E., J.L. Ludke and G.G. Murphy. 1966. Trans. Amer. Fish. Soc. 95: 335.

Gackstatter, J.H. 1966. The uptake from water by several species of fresh water fish of p,p'-DDT, dieldrin, and lindane; their tissue distribution and elimination rate. Ph.D. dissertation, University of North Carolina, Chapel Hill. 140 p.

________ and C.M. Weiss. 1967. Trans. Amer. Fish. Soc. 96: 301.

Grzenda, A.R., W.J. Taylor and D.F. Paris. 1972. Trans. Amer. Fish. Soc. 101: 686.

________. 1971. Trans. Amer. Fish. Soc. 100: 215.

________, D.F. Paris and W.J. Taylor. 1970. Trans. Amer. Fish. Soc. 99: 385.

Giblin, F.J. and E.J. Massaro. 1973. Toxicol. Appl. Pharmacol. 24: 81.

Gorbach, S. and W. Knauf. 1972. In: Environmental Quality and Safety. Vol. 1: 250 (F. Coulston and F. Korte, Eds.). Acad. Press.

Haque, R., V.H. Freed and P.C. Kearney. 1977. See this book.

Holden, A.V. 1962. Amer. Appl. Ecol. 50: 467.

Kenaga, E.E. 1975. In: Environmental Dynamics of Pesticides (R. Haque and V.H. Freed, Eds.). Plenum Press. p. 217-274.

Khan, M.A.Q., F. Korte and J.F. Payne. 1977. See this book.

Khan, H.M., S. Neudorf and M.A.Q. Khan. 1975. Bull. Environ. Contam. Toxicol. 13: 582.

Lee, R.F., R. Sanesheber and G.H. Dobbs. 1972. Marine Biol. 17:201.

Macek, K.J. 1970. In: The Biological Impact of Pesticides

in the Environment. (J.W. Gillette, Ed.). Oregon State University Press. p. 17-21.

Matsumura, F. 1977. See this book.

Meickle, R.W., N.H. Kurihara and C.R. Youngsen. 1972. A bioconcentration study utilizing DDT and mosquitofish: cited by E. Kenaga in Environmental Dynamics of Pesticides. (R. Haque and V.H. Freed, Eds.). Plenum Press. p. 217-274.

Metcalf, R.L. 1977. See this book.

Moore, R., E. Toro and M.A.Q. Khan. 1977. Archiv. Environ. Contam. Toxicol.

Moss, J.A. and D.E. Hathway. 1964. Biochem. J. 91: 384.

Mount, D.I. and G.J. Putnicki. 1966. N. Amer. Wildlife Resources Conf. 31: 177.

Neff, J.M. and J.W. Anderson. 1974. Proc. Nat. Shellfish Assoc. 64:

Premdas, F.H. and J.M. Anderson. 1963. J. Fish. Res. Bd. Canada 20: 827.

Rudd, R.L. 1964. Pesticides and Living Landscape. University of Wisconsin Press. 240 pp.

Schoenthal, N. 1963. Proc. Montana Acad. Sci. 23: 63.

Schoettger, R.A. 1970. Toxicology of Thiodan in several fish and aquatic invertebrates. Investigations in Fish Control, 35, U.S. Depart. of Interior, Bureau of Sport Fisheries and Wildlife. U.S. Govt. Printing Office, Washington, D.C.

Schultz, D.P. and P.D. Herman. 1976. J. Fish. Res. Bd. Canada 33: 1121.

Smith, A.L., R.H. Green and A. Lutz. 1975. J. Fish. Res. Bd. Canada 32: 1297.

Stegeman, J.J. and J.M. Teal. 1973. Marine Biol. 22: 37.

Statham, C.N. and J.J. Lech. 1975. J. Fish. Res. Bd. Canada 32: 315.

Stanton, R.H., H.M. Khan, D.F. Rio and M.A.Q. Khan. 1973. Paper presented at National Meeting of the American Chemical Society, A 29, 1973, Chicago, Illinois.

Tooby, T.E. and F.J. Durbin. 1975. Environ. Poll. 8: 79.

Woodwell, G.M., C.F. Wurster and P.I. Issacson. 1967. Science 156: 821.

ACKNOWLEDGEMENTS

Supported by a U.S.P.H.S. Grant (ES - 01479) from the National Institute of Environmental Health Sciences.

MODEL ECOSYSTEM STUDIES OF BIOCONCENTRATION AND BIODEGRADATION OF PESTICIDES

Robert L. Metcalf

INTRODUCTION

Laboratory model ecosystems of microcosms are now well-established tools for investigating the environmental toxicology of pesticides, using radiolabeled molecules under standardized laboratory conditions. Laboratory model ecosystems are potentially almost as diversified as the natural environment that is being modeled and in practice have ranged in complexity from petri dishes containing soil microflora and flasks containing aquatic organisms to very complex model streams and terrestrial chambers often highly instrumented and designed for computer analysis. All of these model ecosystems are designed and operated to study (a) chemical degradation pathways, (b) transport, fate, and accumulation in living organisms, (c) toxicological effects on various organisms, (d) biochemical, physiological, and ecological parameters of environmental toxicology.

In short model ecosystems are designed as "early warning systems" to screen pesticides (and other chemicals) for adverse environmental effects. Their use affords quantitative as well as qualitative estimates of the environmental impact of chemicals. The model ecosystem approach is essentially comparative and pesticides are evaluated vis-a-vis certain well established compounds such as DDT, dieldrin, 2,4-D etc. whose real world environmental toxicology has been studied in great detail. A variety of structural modifications of a potentially useful pesticide nucleus, e.g. triazine, diphenyl ethane, pyrethroid, or growth regulator may be compared in detail to permit selection of derivatives which optimize target-site effectiveness with appropriate environmental safety.

This paper will present results obtained using a terrestrial-aquatic model ecosystem with several food chains, to evaluate a variety of environmental problems related to the development and usage of pesticides.

METHODOLOGY

The model ecosystem methodology has been described in detail (Metcalf et al. 1971, Metcalf 1974) and will be but briefly discussed here. The basic unit is a glass aquarium 25 x 30 x 46 cm, incorporating a sloping shelf of 15 kg of washed white quartz sand, bisected by a lake of seven liters of standard reference water (Figure 1). On the terrestrial shelf are grown 50 *Sorghum vulgare* plants to comprise a "farm". The aquatic or "lake" portion contains a complement of microorganisms, plankton, 30 *Daphnia magna*, 10 *Physa* sp. snails, and a clump of alga *Oedogonium cardiacum*. After the model ecosystem has equilibrated for 26 days in an environmental plant growth chamber of 26°C with 12 hour daylight exposure to 5000 ft. candles (54,000 lux), 200 *Culex pipiens* mosquito larvae are added, and after 30 days, three *Gambusia affinis* fish. The experiment is customarily terminated at 33 days after treatment but can be extended to measure duration of toxic effects on the organisms. The unit is covered with a plexiglass top incorporating a screen wire portion over the terrestrial end. This top retards evaporation and yet permits photochemical effects and air exchange.

In operation, a pesticide radiolabeled with ^{14}C, ^{13}H, ^{32}P, or ^{35}S is applied quantitatively to the *Sorghum* leaves from acetone solution using a micropipette. The dosage used is 1.0 to 5.0 mg which on the "farm" portion is equivalent to about 0.2 to 1.0 kg per hc. Alternatively, the radiolabeled compound can be incorporated into the sand by injection from a microsyringe, applied in soil which is incorporated in the sand or applied as seed dressing or granular. After application, 10 *Estimene acrea*, salt marsh caterpillar

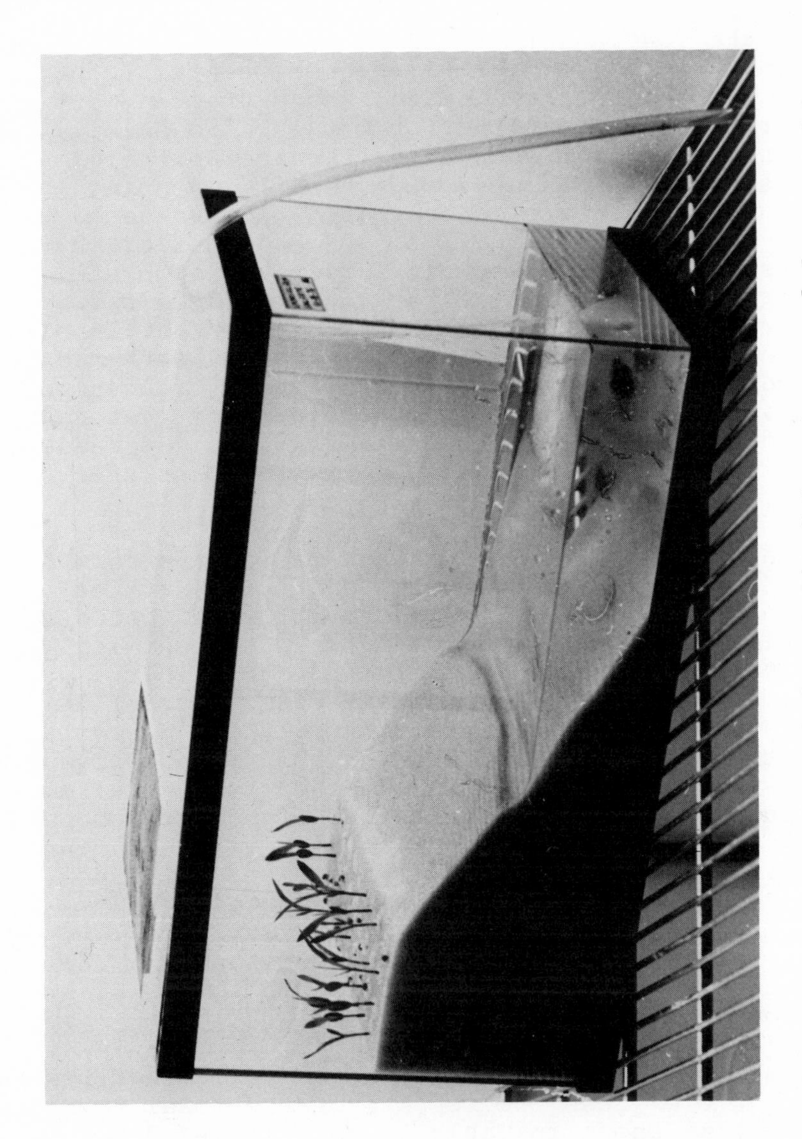

Figure 1. Terrestrial-aquatic laboratory model ecosystem.

larvae, are added to the system to consume the *Sorghum* plants and initiate the dispersal of the radiolabeled pesticide and its degradation products.

A. Evaluation of Results

At the conclusion of the 33 day experiment, the water and weighed samples of the components of the model ecosystem are homogenized and extracted with acetonitrile and evaluated for total radioactivity by liquid scintillation counting. The acetonitrile extracts of plant tissues, alga, snail, daphnia, mosquito, and fish are concentrated and evaluated by thin layer chromatography (TLC) and radioautography. The tissue residues after extraction are combusted to $^{14}CO_2$ or $^{3}H_2O$ to determine the unextractable radioactivity. The nature and amounts of the radiolabeled products are determined by R_f as compared with model compounds, by microchemical tests, and by removal of the radiolabeled areas from the thin-layer plates followed by quantitative determination of the radiolabel and by possible identification by mass spectrometry.

Measurements and Observations which are commonly made in the model ecosystem include: (1) rate of contamination of the water phase, (2) toxic effects on the various organisms, (3) changes in the chemical constitution of the radiolabeled products in the water and living organisms, (4) percentages of parent compound in the extractable radioactivity and percentages of unextractable radioactivity in the various organisms and substrates, (5) determinations of ecological magnification (E.M.) or ratio of amount of parent compound or specific degradation products in organisms/amount in water, (6) determinations of biodegradability index (B.I.) or amount of polar radioactivity in organism/amount of non-polar radioactivity, (7) percent biomass recovery in the various organisms, and (8) observations on possible toxic effects to the plants and animals in the model ecosystem.

Summary Tabulation. A standard format has been developed to express the results of the model ecosystem experiment as shown in Table 1 (Metcalf and Sanborn 1975). The quantitative data from the water and various organisms of the system is expressed in parts per million (ppm) of total extractable radioactivity in equivalents of the parent pesticide as determined

Table 1. Model Ecosystem Study with Three Radiolabeled Preparations of Diflubenzuron[1/]

		ppm - diflubenzuron equivalents								
		H_2O			Culex (mosquito)			Gambusia (fish)		
	R_f[2/]	A	B	C	A	B	C	A	B	C
Extractable-total		0.02356	0.06909	0.00615	5.2455	13.3614	1.9320	1.6444	6.0701	0.7730
I	0.90				0.7201					
$(CH_3)_2NC_6H_4Cl$	0.83		0.0034							
II	0.77	0.00025		0.00005						
$C_6H_3F_2CONHCONHC_6H_4Cl$	0.70	0.0057	0.0220	0.00168	4.4225	13.1369	1.8466	0.1097	0.3193	0.1350
$C_6H_3F_2COOH$	0.52	0.0018		0.00037			0.0777	0.1644		0.0226
$H_2NC_6H_4Cl$	0.50		0.0078						0.3193	
$C_6H_3FCONH_2$	0.45	0.00031		0.00008						0.0189
III	0.43		0.00055							
IV	0.38	0.00020								
V	0.36		0.00061							
$CH_3CONHC_6H_4Cl$	0.33		0.00056						0.0297	
VI	0.26		0.00022							
VII	0.20		0.00015							
$H_2NCONHC_6H_4Cl$	0.12		0.0072			0.2071			0.1263	
Polar	0.0	0.0060	0.0078	0.00172	0.1028	0.0174	0.0077	1.3703	5.2755	0.5965
Unextractable		0.0093	0.0188	0.00225	1.9340	0.3436	0.0251	0.8227	0.8469	0.1328

1/ TLC with benzene-dioxane-acetic acid (90:30:1)
2/ Metcalf et al. (1975)

from the acetonitrile extracts and in ppm of unextractable radioactivity as determined by total combustion analysis to $^{14}CO_2$ or $^{3}H_2O$. The results from TLC determinations of the radiolabeled degradation products are calculated in terms of percent of total extractable radioactivity and are tabulated as ppm equivalents ranging from highest to lowest R_f on the radioautograms of the TLC plates (Table 1). Thus the non-polar compounds appear at the top of the table and the polar compounds at the bottom. This sort of tabulation simplifies calculations of ecological magnification or E.M. The Summary Tabulation also facilitates the calculation of percentage parent compound in the organisms (measuring persistence) and percentage of unextractable radioactivity (measuring total degradation and resynthesis), in the water and organisms.

The data of Table 1 was obtained from three model ecosystems each treated with a different radiolabeled preparation of the insect growth regulator diflubenzuron (Dimilin®): A with benzoyl $^{14}C=O$, B with ^{14}C phenyl ring, and C with ^{3}H benzyl ring. Comparisons of the results of the three labels are very important in characterizing the environmental degradation of the molecule. For example, cleavage of diflubenzuron at benzoyl-CO-NH produces benzoic acid and benzamide fragments labeled in A and C and phenyl urea fragment labeled with B. Cleavage at phenyl-NH-CO produces aniline fragments labeled in B. Unknown degradation products are characterized by type of label as relating to benzyl ring or phenyl ring. Decarboxylation products would be labeled only with C.

Reliability and Reproducibility of Data. The three replicate determinations of diflubenzuron shown in Table 1, also give a good indication of the reproducibility of results. As seen in the Table, the level of water contamination was somewhat variable for the three systems, as it depends upon the feeding and excreting habits of the salt marsh caterpillar larvae. However, this does not appreciably affect the quantitative information obtained. For example, the E.M. values for the three systems were fish A 19, B 14, and C 80 and those for the mosquito larvae were A 779, B 596, and C 1099. The differences between fish and mosquito reflect both the enhanced detoxication in the fish and the particular affinity of this insect growth regulator for the site of action in the insect cuticle. The B.I. values for the three systems were fish A

4.99, B 6.64, and C 3.38 and mosquito A 0.020, B 0.001, and C 0.004. The large differences between fish and mosquito indicate the ready detoxication of diflubenzuron in fish and its pronounced stability in the mosquito larvae. It should be pointed out that the labeling position should have some effect on B.I. values depending upon the relative polarity of the labeled fragments produced in degradation.

B. Alternative Model Systems

Several variations of the standard model ecosystem have also been used in this laboratory. The aquatic model ecosystem (Lu and Metcalf 1975) uses three-liter 3-necked flask with 2 liters of standard reference water containing the same combination of aquatic organisms described above. The flask is fitted with condensor and traps for organic vapors and $^{14}CO_2$. This system operated over three days has been particularly valuable for determining initial rates of biomagnification and comparative metabolic pathways for degradation.

The feed lot model ecosystem (Coats et al. 1977) is used to evaluate veterinary drugs and pesticides administered in feed, or directly to mice and chicks caged above the terrestrial-aquatic interface of the standard model ecosystem. Thus the radiolabeled "drugs" and their degradation products can be followed through the animals and from their excreta throughout the organisms of the terrestrial and aquatic phases.

The "rice-paddy" model ecosystem (Lee et al. 1976) incorporates rice *Oryza sativa*, grown on the terrestrial shelf of essentially the standard model ecosystem with a higher water table. It is used to model the unique rice-paddy environment and to study the interactions of insecticides, herbicides, and fungicides applied pre- and post-emergent to the rice paddy and their effects on the living environment.

RESULTS

During the past 10 years our laboratory has evaluated nearly 100 radiolabeled pesticides by the model ecosystem technology described above. The initial study (Metcalf et al. 1971) discussed the comparative environmental behavior of DDT, DDD, DDE, and methoxychlor. Kapoor et al (1970, 1972, 1973) discussed the model ecosystem behavior of 8 DDT analogues covering a wide range of biodegradability. The environmental fate of aldrin, dieldrin, endrin, lindane, mirex, and hexachlorobenzene was reported by Metcalf et al. (1973) and these studies were extended to chlordene, heptachlor, and heptachlor epoxide (Lu et al. 1975), and chlordane and toxaphene (Sanborn et al. 1976). The comparative fate of three dichlorophenyl nitrophenyl ether herbicides was studied by Lee et al. (1976). Metcalf and Sanborn (1975) summarized in detail, model ecosystem studies with 49 radiolabeled pesticides. This review very briefly summarizes aspects of these and other model ecosystem studies on pesticides.

A. Screening of New Candidate Insecticides

The laboratory model ecosystem has been used in this laboratory to screen a variety of analogues of DDT for environmental degradability. These compounds were synthesized as part of a long term project to evaluate the principles of biodegradability. The basic methodology involved systematic study of the DDT molecule by replacing the environmentally stable C-Cl bonds with other groups of suitable size, shape, and polarity that could serve as degradophores by acting as substrates for the mixed function oxidase enzymes widely distributed in living organisms (Metcalf et al. 1971, Coats et al. 1974). The action of the mixed function oxidase detoxication enzymes on molecular moieties which can serve as substrates was shown to result in substantial changes in the polarity of the molecule so that the degradation products were excreted rather than stored in lipids as was DDT and its chief degradation product DDE. For example methoxychlor or 2,2-bis(*p*-methoxyphenyl)-1,1,1-trichloroethane, water solubility 0.62 ppm, is converted *in vivo* by *O*-demethylation to 2,2-bis-(*p*-hydroxyphenyl)-1,1,1-trichloroethane, water solubility 76 ppm; and methylchlor or 2,2-bis-(*p*-methylphenyl)-1,1,1-trichloroethane, water solubility 2.21 ppm, is converted *in vivo* by side chain oxidation to 2,2-bis-(*p*-carboxyphenyl)1,1,1-trichloroethane, water solubility 50 ppm (Kapoor et al. 1970, 1972, 1973).

A summary of model ecosystem data for a number of DDT analogues with degradophores incorporated into aromatic and aliphatic moieties of the molecule is shown in Table 2. This type of information has considerable value in the demonstration of biodegradability and environmental compatibility for specific molecular modifications of the desired active nucleus.

B. Suitability of Insecticides for Vector Control

The World Health Organization is developing a new program in the Volta River basin of West Africa to control onchocerciasis. This human disease caused by the filarian parasite *Onchocerca volvulus* affects more than 20 million persons of whom 20 percent or more may become blind. Onchocerciasis is transmitted between humans by the black flies *Simulium* spp. which breed as larvae in running water of W. Africa streams and rivers. WHO is attacking the disease by controlling *Simulium* breeding through larviciding. The environmental effects of larvicides on non-target organisms are very important in the selection of suitable insecticides, particularly because of the importance of fishing in rivers, reservoirs, and estuaries. Laboratory screening selected a number of candidate larvicides of which chlorpyrifos *O,O*-diethyl O-(3,5,6-trichloro-2-pyridyl) phosphorothionate was among the most effective. However, the corresponding *O,O*-dimethyl ester was almost as effective as a larvicide and much less hazardous to man and higher animals. As part of the selection criteria, the relative environmental degradability of the two esters was compared in laboratory model ecosystems using 3,5,6-trichloro-2-pyridyl ^{14}C-2,6-labeled phosphorothionate esters (Metcalf 1974). The results as shown below provided an excellent demonstration of the value of the model ecosystem as a predictive early warning system:

	chloropyrifos	chloropyrifos methyl
parent compound in fish, ppm	0.0352	0.0076
parent compound, % extractable	49.50	20.71
unextractable ^{14}C, %	23.9	52.2
ecological magnification (E.M.)	314	95
biodegradability index (B.I.)	1.02	3.95

Chloropyrifos methyl is evidently substantially more biodegradable and less accumulative in the fish *Gambusia* than chloropyrifos. The higher value for percent unextractable ^{14}C indicates the greater lability of chloropyrifos methyl in the organisms and

Table 2. Model Ecosystem Characterization of Biodegradability of DDT Analogues [1/]

R^1	R^2	R^3	fish E.M.	B.I.
Cl	Cl	CCl_3	84,500	0.015
Cl	Cl	$HCCl_2$	83,500	0.054
CH_3O	CH_3O	CCl_3	1,545	0.94
CH_3	CH_3	CCl_3	140	7.14
CH_3S	CH_3S	CCl_3	5.5	47
Cl	CH_3	CCl_3	1,400	3.43
CH_3	C_2H_5O	CCl_3	400	1.20
CH_3O	CH_3S	CCl_3	310	2.75
CH_3O	CH_3O	$C(CH_3)_3$	1,636	1.04
Cl	Cl	$HC(CH_3)NO_2$	112	3.27

[1/] Data from Metcalf et al. (1971), Kapoor et al. (1973), Hirwe et al. (1975) and Coats et al. (1974).

the resynthesis of the ^{14}C into the products of endogenous metabolism. On the basis of lower toxicity and higher biodegradability chlorpyrifos methyl was selected as the more suitable larvicide for *Simulium* control. Comparison of the E.M. and B.I. values for these organophosphorous insecticides with those for DDT and methoxychlor (Table 2) indicates the importance of the increased biodegradability of these organophosphorus insecticides in preventing the environmental pollution of fish and other aquatic organisms.

A similar model ecosystem study made of temephos or Abate^{R} which is also an effective compound for control of *Simulium* larvae, showed the value of the model ecosystem in demonstrating environmental safety of pesticides (Metcalf and Sanborn 1975). From the ^{3}H temephos, 12 oxidative and hydrolytic products were ioslated from water and alga and snail. No identifiable products were isolated from the fish which contained ^{3}H equivalent to only 0.00099 ppm. Temephos was found in alga at 0.00195 ppm and in snail at 0.01876. This study showed the exceptional biodegradability of temephos with its fish E.M. value of 0 and B.I. of ∞ suggesting that its use as a replacement for DDT, fish E.M. 84,500 and B.I. 0.015, would greatly lessen the environmental impact of larviciding for *Simulium*.

C. Degradative Pathways of Pesticides

Knowledge of the chemical pathways by which pesticides are degraded in the environment is essential to understand their environmental impact and for the registration and licensing of new products. The importance of such studies is gauged by a recent critique of 32 widely used pesticides registered with EPA stating that data on soil degradation were lacking for 11, on water degradation for 17, on microbial degradation for 16, and in leaching and run-off for 24 (Chemical & Engineering News 1976). Laboratory model ecosystem studies conducted with radiolabeled pesticides provide a simple and logical way to develop information on their environmental fate and degradation not only in the overall system but also on a comparative basis in water, soil, plant, and individual components of food chains.

A recent model ecosystem study of the fate of the new insect growth regulator diflubenzuron or 1-(2',6'-difluorobenzoyl)-3-(4'-chlorophenyl) urea provided a good example of the wealth of information to be secured from such studies (Metcalf et al. 1975). The type of information obtained with three radiolabeled preparations furnished by the manufacturer is shown in Table 1. The availability of the several radiolabels made it possible to follow both aromatic portions of the molecule following biochemical and photolytic cleavage of diflubenzuron at the peptide-type links $C_6H_3F_2$CO-NH-CO-NHC_6H_4Cl. The major products identified from A and C radiolabels (Table 1) were 2,6-difluorobenzamide and 2,6-difluorobenzoic acid, and from the B radiolabel were *p*-chlorophenyl-urea and *p*-chloroaniline. In analogy with aniline (Lu and Metcalf 1975) it was found that *p*-chloroaniline was transformed into the *N*-methyl *N*,*N*-dimethyl, and *N*-acetyl derivatives.

Other examples of elucidation of environmental degradative pathways of pesticides in model ecosystems include methoxychlor (Kapoor et al. 1970), methylchlor (Kapoor et al. 1972), prolan (Hirwe et al. 1975), dianisyl neopentane (Coats et al. 1974), aldrin and dieldrin (Metcalf et al. 1973), heptachlor (Lu et al. 1975), and diphenyl ether herbicides (Lee et al. 1976).

D. Comparative Metabolism

Model ecosystem studies with radiolabeled pesticides provide useful data about the comparative metabolism of the pesticides in a variety of organisms of several different phyla. This has become an integral part of the evaluation with the model aquatic ecosystems (Lu and Metcalf 1975) which has the same complement of organisms as the terrestrial-aquatic system. Studies with ^{14}C hexachlorocyclopentadiene, chlordene, heptachlor, and heptachlor epoxide provided a good example (Lu et al. 1975). These four compounds represent stages in the environmental history of an insecticide from raw material to intermediate and principal impurity (chlordene) to product (heptachlor) and to environmental pollutant (heptachlor epoxide). Chlordene was oxidized to chlordene-2,3-epoxide and hydrolyzed to 1-hydroxy-2,3-epoxychlordene. Heptachlor was epoxidized to form heptachlor-2,3-epoxide which formed 22% of the extractable ^{14}C in alga, 37% in snail, 49% in mosquito, and 79% in fish. Heptachlor was also hydrolyzed at C_1 to 1-hydroxychlordene, but heptachlor epoxide without the activating allylic structure (CH=CHCHCl) was relatively inert, and was stored in all of the organisms, 91% of the

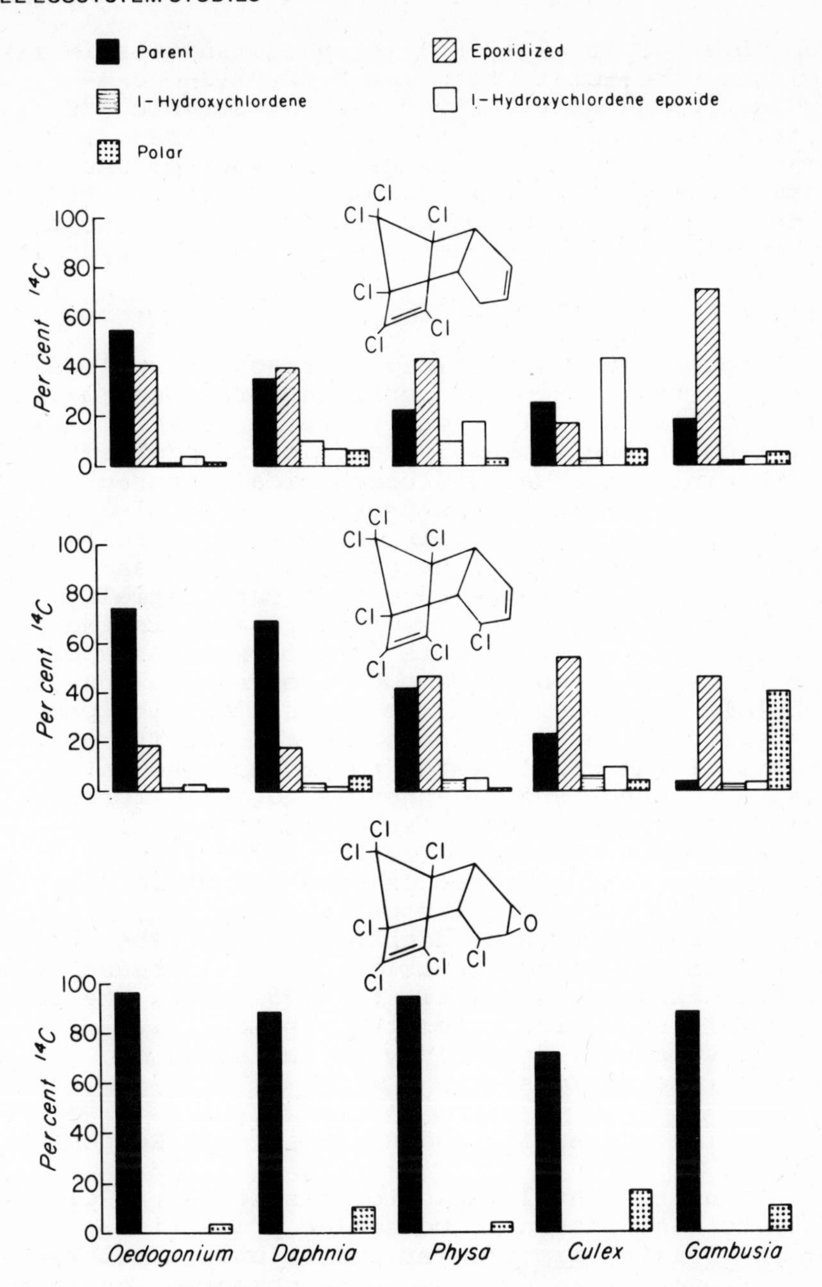

Figure 2. Relative detoxication capacities of key organisms of model aquatic ecosystem treated with ^{14}C cyclodienes. From Lu et al. 1975. Copyright by the American Chemical Society, reprinted by permission.

extractable ^{14}C in alga, 82% in snail, and 69% in fish. Hydrolysis of C_1-Cl to form 1-hydroxy-2,3-epoxy-chlordene formed only 3.5% of the extractable ^{14}C in alga, 8.7% in snail, and 19% in fish. These data (summarized in Figure 2) demonstrate the reasons for the environmental problems encountered with heptachlor, e.g. the *in vivo* formation of the very stable heptachlor epoxide which does not readily undergo hydrolysis.

E. Pesticide Interactions

Although various pesticides are often applied together in tank mixes and persistent residues may exist together in soil or water, there is very little information on the immensely complicated problems of their possible chemical and biochemical interactions. Laboratory model ecosystems provide a suitable way to study these. For example, inhibitors of microsomal oxidases such a piperonyl butoxide are used as insecticide synergists because they substantially alter the detoxication rates of pesticides in the living insects. A study with ^{3}H-methoxychlor applied to the terrestrial-aquatic model ecosystem alone and with 5-fold piperonyl butoxide showed that the presence of the synergist severely inhibited the ability of the organisms of the model system to produce the water partitioning mono- and bis-phenols normally formed by *O*-demethylation. As a result the organisms in the model ecosystem treated with piperonyl butoxide contained two to four times as much intact methoxychlor along with increased storage of methoxychlor ethylene and methoxychlor dichloroethane. The net result of the piperonyl butoxide was to increase E.M. by 3.3-fold and decrease B.I. to 0.25-fold. (Metcalf 1974). This study indicates the applicability of the model ecosystem technology to the study of pesticide interactions.

F. Molecular Properties and Environmental Response

The environmental impact of pesticides is clearly related to such intrinsic molecular properties as water solubility, lipid/water partitioning, polarity, and electronic distribution. Development of quantitative relationships between these properties and environmental parameters that can be determined from model ecosystem investigations, e.g. bioconcentration (ecological magnification), biodegradability index, and percent unextractable radioactivity, would provide "early warnings" of the pollutant properties of new or proposed environmental contaminants. Using simple radiolabeled organic molecules, Lu and Metcalf (1975) showed substantial correlation between ecological magnification (E.M.) for model ecosystem studies with water insolubility and with octanol/H_2O partition coefficient. These studies were extended to more than 30 pesticides by Metcalf and Sanborn (1975) who found a highly significant correlation between log E.M. and log water insolubility (Figure 3) and between log E.M. and log percent unextractable radioactivity. It has been suggested (Metcalf and Sanborn 1975) that pesticides with water solubilities <0.5 ppm are likely to show objectionable biomagnification and those with solubilities >50 ppm are unlikely to be biomagnified. The pesticides in the in-between range should be regarded with caution. This relationship is clearly the result of a highly significant correlation between water solubility and lipid/water partitioning and when accurate values for octanol/H_2O partition coefficients for a large number of pesticides are available even more useful relationships can be developed.

CONCLUSIONS

Model ecosystem technology, with the use of radiolabeled molecules, is the most informative and convenient tool for studying the fate and environmental effects of pesticides. The technique is rapid and inexpensive and can be used in virtually any laboratory. Through its expanded use it will be possible to fill many of the lacunae in our knowledge of the fate of pesticides in the environment, to quantify the principles of biodegradability, and to design new pesticide molecules for greater environmental compatability.

ACKNOWLEDGEMENTS

The writer expresses his great appreciation to the many coworkers named in the bibliography who have participated in model ecosystem development and evaluation. The research described was supported by grants from the Herman Frasch Foundation, The Rockefeller Foundation, the National Science Foundation (Grant ESR 74-22760), the Environmental Protection Agency (Grant R-80-3249), the World Health Organization,

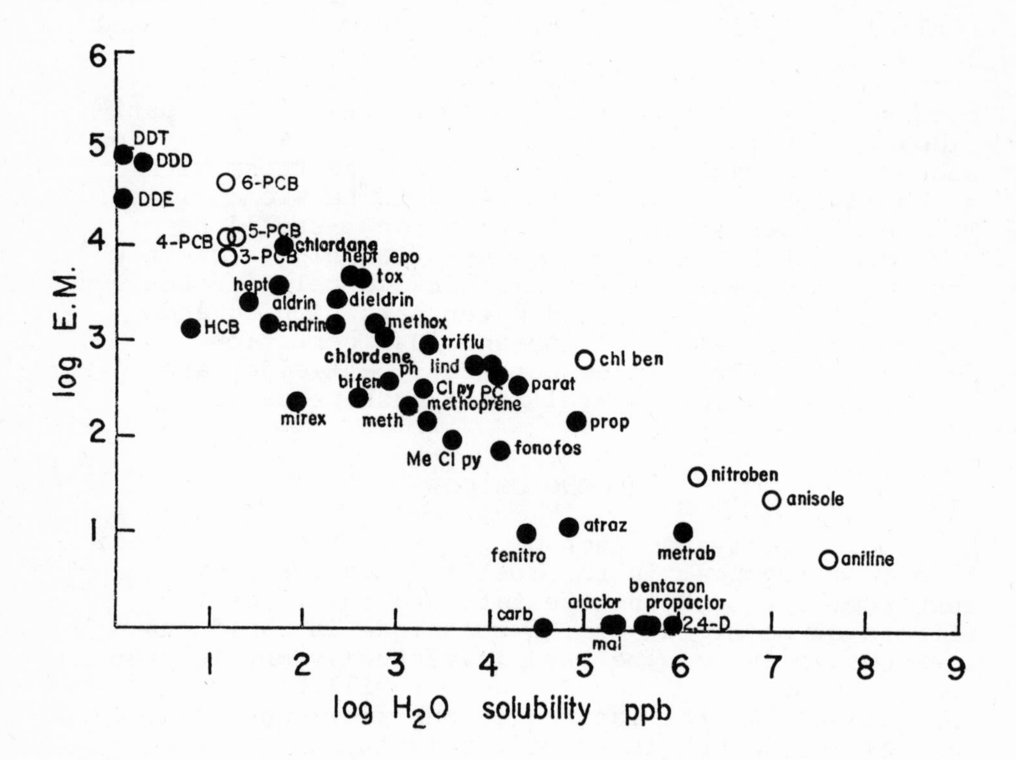

Figure 3. Relationship between water solubility of pesticides and ecological magnification in mosquito fish of terrestrial-aquatic model ecosystem. After Metcalf and Sanborn, 1975.

The U. S. Agency for International Development (Contract AIC/ta-C-1228), and the Department of Interior through the Illinois Water Resources Center (Project B-050 Illinois).

REFERENCES

Chemical & Engineering News, Feb. 16, pp 19 (1976).

Coats, J. R., R. L. Metcalf, Po-Yung Lu. 1977. Environ. Health Perspectives. In Press.

Hirwe, A. S., R. L. Metcalf, Po-Yung Lu, and Li-Chun Chio. 1975. Pesticide Biochem. Physiol. 5: 65.

Kapoor, I. P., R. L. Metcalf, A. S. Hirwe, J. R. Coats, M. S. Khalsa. 1973. J. Agr. Food Chem. 21: 310.

Kapoor, I. P., R. L. Metcalf, A. S. Hirwe, Po-Yung Lu, J. R. Coats, and R. F. Nystrom. 1972. J. Agr. Food Chem. 20: 1.

Kapoor, I. P., R. L. Metcalf, R. F. Nystrom, and G. K. Sangha. 1970. J. Agr. Food Chem. 18: 1145.

Lee, An-Horng, Po-Yung Lu, R. L. Metcalf, and Err-Lieh Hsu. 1976. J. Environ. Quality 5: 482.

Lu, Po-Yung and R. L. Metcalf. 1975. Environ. Health Perspectives 10: 269.

Lu, Po-Yung, R. L. Metcalf, A. S. Hirwe and J. W. Williams. 1975. J. Agr. Food Chem. 23: 967.

Metcalf, R. L. 1974. Essays Toxicology 5: 17.

Metcalf, R. L., I. P. Kapoor, Po-Yung Lu, C. Schuth, and P. Sherman. 1973. Environ. Health Perspectives 4: 35.

Metcalf, R. L., Po-Yung Lu, and S. Bowlus. 1975. J. Agr. Food Chem. 23: 359.

Metcalf, R. L. and J. R. Sanborn. 1975. Bull. Illinois Natural History Survey 31, Art. 9, pp 381-436.

Metcalf, R. L., G. K. Sangha, and I. P. Kapoor. 1971. Environ. Sci. Technol. 5: 709.

Sanborn, J. R., R. L. Metcalf, W. N. Bruce, and Po-Yung Lu. 1976. Environ. Entomology 5: 533.

THE USE OF LABORATORY DATA TO PREDICT THE DISTRIBUTION OF CHLORPYRIFOS IN A FISH POND

W. Brock Neely and Gary E. Blau

ABSTRACT

In the commercial development of any new chemical, a great deal of laboratory data are generated. One use of these data is to make predictions of the impact that will result from the introduction of the compound into the environment during manufacture, use or disposal. This investigation demonstrates how such data can be used to make these predictions in a more quantitative manner than has been reported in the past. The credibility of the predictions is strengthened by matching the results with an actual field study.

INTRODUCTION

Knowledge of the environmental fate of chemicals has always been important but with increasing awareness of their potential environmental effects, such studies have taken on even greater importance. Society wants to know, before any action is taken, both the immediate and long-term consequences of that action in all phases of the environment. Scientists must continue to improve their ability to make predictions based on the best data available. From an environmental point of view, it is necessary to improve our predictive ability by comparing laboratory data with actual environmental behavior. In the past, all too often products have been in use for several years before their effects have been seen.

Classic examples of such hindsight are DDT and PCB. Today our qualitative predictive abilities are much improved, but quantitative correlations are still lacking.

Many environmentalists have been very skeptical about the ability to translate laboratory data to actual field observations. It is for this reason that the Agricultural Industry has relied very heavily on field studies to verify predictions that have been made in the laboratory. Unfortunately, many field studies are conducted in a manner that makes quantitative analysis very difficult if not impossible. Recently, Blau & Neely[1] characterized the distribution of Chlorpyrifos [O,O-Diethyl-O-(3,5,6-trichloro-2-pyridyl) phosphorothioate] in a laboratory ecosystem. Macek et al.[2] added the same material to several 0.1 acre fish ponds in Missouri. In both studies the level of Chlorpyrifos was determined in the water and fish. In this paper we will examine the ability of the laboratory data to predict these field data and attempt a generalization of this concept.

THE FIELD STUDY

In order to set the stage for this presentation a description of the fish ponds is desirable. This can be accomplished by quoting directly from the article[2]: "The investigation was conducted at the Fish Pesticide Research Laboratory near Columbia, Missouri during the summer of 1969. We utilized six 0.1 acre rectangular earthern ponds containing a measured volume of water. These ponds have a mean depth of 0.25 m at the shallow end and 2.0 m at the deep end. Aquatic vegetation present consisted primarily of cattail, arrowhead, pondweed and filamentous green algae." The ponds were stocked with 275 largemouth bass (17.1 ± 2.7 g) and 275 bluegills (14.1 ± 3.9 g). Two of the ponds were treated with Chlorpyrifos at the rate of 0.05 lb/acre, two were treated at the rate of 0.1 lb/acre and the remaining two were held as controls. The Chlorpyrifos was dissolved in 20% acetone in water and sprayed over the surface of the ponds with a conventional garden sprayer.

After applying the Chlorpyrifos, water samples and fish were collected and analyzed for the insecticide on days 1, 3, 7, 14, and 28. The pond parameters in metric units are shown in Table I.

TABLE I

POND PARAMETERS FOR FIELD STUDY OF CHLORPYRIFOS

Volume (V)	3.59×10^{8} ml
Area (A)	4.04×10^{6} cm^2
Average depth (D)	88.8 cm
Total wt of fish (F)	8580 g
Total wt of soil (S)	1.49×10^{7} g*
Initial concentration of Chlorpyrifos at 0.05 lb/acre	5.75 ppb†

*This is based on the conversion factor of 3.7 g soil/cm^2 for a 2.5 cm layer.
†The mean temperature on the day of application was 25°C.

FISH POND MODEL

In analyzing the ponds, it was necessary to construct a model which would describe the movement and distribution of the chemical in this system. The model used is shown in Figure 1. Assuming idealized volume and first order rate processes, a material balance for the water phase yields equation 1. The building of such a compartmental model and the limitation in their use have been described in previous reports[1,3].

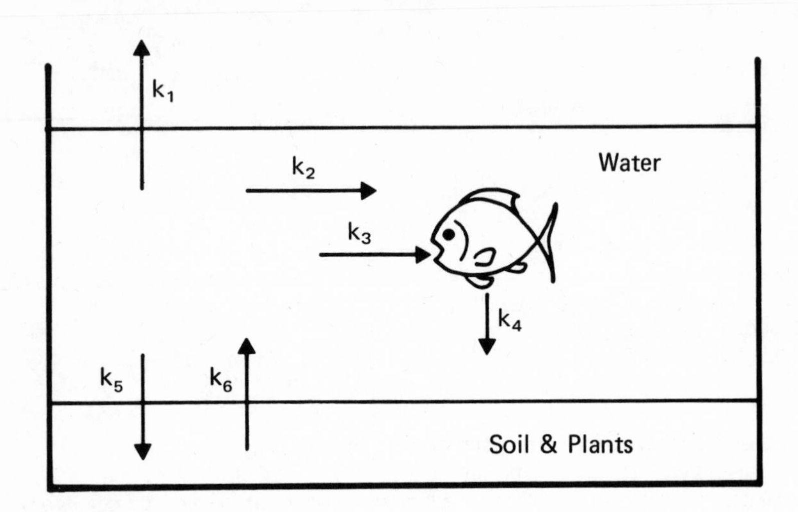

Figure 1. Schematic representation of the movement of Chlorpyrifos in the fish ponds.

TABLE II

CHEMICAL AND PHYSICAL PROPERTIES OF CHLORPYRIFOS INSECTICIDE

Water solubility (ppm)	2.0 at 25°C
Vapor pressure (mm Hg)	1.87×10^{-5} at 25°C
Molecular wt	350.5
Partition coefficient (octanol/water)	66,000
Velocity constant (water → air) (cm/hr)	0.12
Half-life in water at a pH of 7.0 (days)	3

$$\frac{VdCw}{dt} = -k_1AC_w-k_2VC_w-k_3FC_w+k_4FC_f-k_5SC_w+k_6SC_s \quad (1)$$

where V = volume (ml)
k_1 = rate const. for evap. (cm/hour)
k_2 = " " " hydrolysis (1/hour)
k_3 = " " " uptake by fish (ml/g fish/hour)
k_4 = " " " excretion by fish (1/hour)
k_5 = " " " uptake by soil (ml/g soil/hour)
k_6 = " " " desorption by soil (1/hour)
C_w, C_f, C_s = conc. in water, fish and soil (gram/gram)
A = area of pond (sq cm)
F = amount of fish (g)
S = wt of soil (g)

Assuming fish may be represented as a single compartment the following material balance is obtained.

$$F\frac{dC_f}{dt} = k_3FC_w - k_4FC_f \quad (2)$$

Estimating the Rate Constants

1. The rate constant for evaporation (k_1) was estimated by the Liss and Slater technique[4] as modified by Neely[5]. This method allows a calculation to be made from a knowledge of the vapor pressure and solubility. This estimate along with other key data is shown in Table II.

2. From Table II a half-life of 3 days at pH 7.0 for the insecticide in water is recorded[6]. The pH of the pond according to Macek et al.[1] was the range of 6.8-8.4. Accordingly, the value for k_2 is equal to 9.6×10^{-3} hours.

3. The remaining rate constants were derived from the laboratory ecosystem study of Smith et al.[7] as analyzed by Blau et al.[1,3] The estimated rate constants were based on total C^{14} activity and consequently must be converted to a concentration basis to reflect the parent compound as measured by Macek[2]. In Smith's study[7] he used 45 g of fish in 6 gallons (22,710 ml) of water. The soil in the aquarium was 5 cm deep and the area of the aquarium was approximately 1,548 cm^2 (20" x 12" aquarium).

Using the conversion factor of 3.7 g soil/cm^2 for a 2.5 cm layer, a weight of 11,455 g of soil is estimated. In the analysis of Smith's aquarium, the model shown in Figure 2 was used. This figure also indicates the optimum parameter values for the various rate constants[3].

In examining the material balance equation it is seen that k_3 must be in units of ml/g fish/hr and k_5 must be units of ml/g soil/hr to make the equation dimensionally correct. This conversion is as follows:

$$k_3 = \frac{k'_3 x 22{,}710}{45} = 55.5 \text{ ml/g fish/hour}$$

$$k_5 = \frac{k'_5 x 22{,}710}{11{,}455} = 0.67 \text{ ml/g soil/hour}$$

k_4 and k_6 may be used directly as k'_4 and k'_6 from Figure 2.

SIMULATION EXPERIMENTS

Simulation 1

In solving equations 1 and 2 on the computer, the following must be noted. In the original modeling of the aquarium[1], it was shown that k_4 characterizes the rate of elimination of some metabolite (probably a pyridinol) from the fish. Consequently in equation 1, k_4 will be zero since it is the parent or intact Chlorpyrifos molecule that is of interest. The results of the simulation modeling are shown in Figure 3 where the actual values as measured by Macek[2] are also shown. The percentage of Chlorpyrifos in the various compartments are given in Table III.

Simulation 2

From the known fish toxicity of Chlorpyrifos[2] (96 hour TL_{50} to bluegills at 18°C was found to be 3.6 ppb with a 95% confidence limit of 1.6-4.1) it can be seen from Figure 3 that some toxicity at this predicted water concentration would be expected. Macek et al.[2] did note some fish kill in their actual experiment. There are several ways to reduce the fish toxicity, a) reduce the initial dose, b) apply to a high organic soil where fast initial absorption will successfully

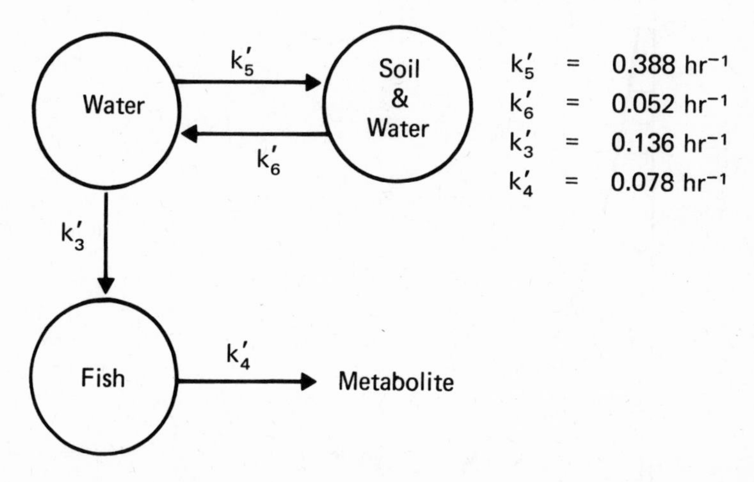

Figure 2. Compartmental model and rate constants for the movement of Chlorpyrifos (Reference 3).

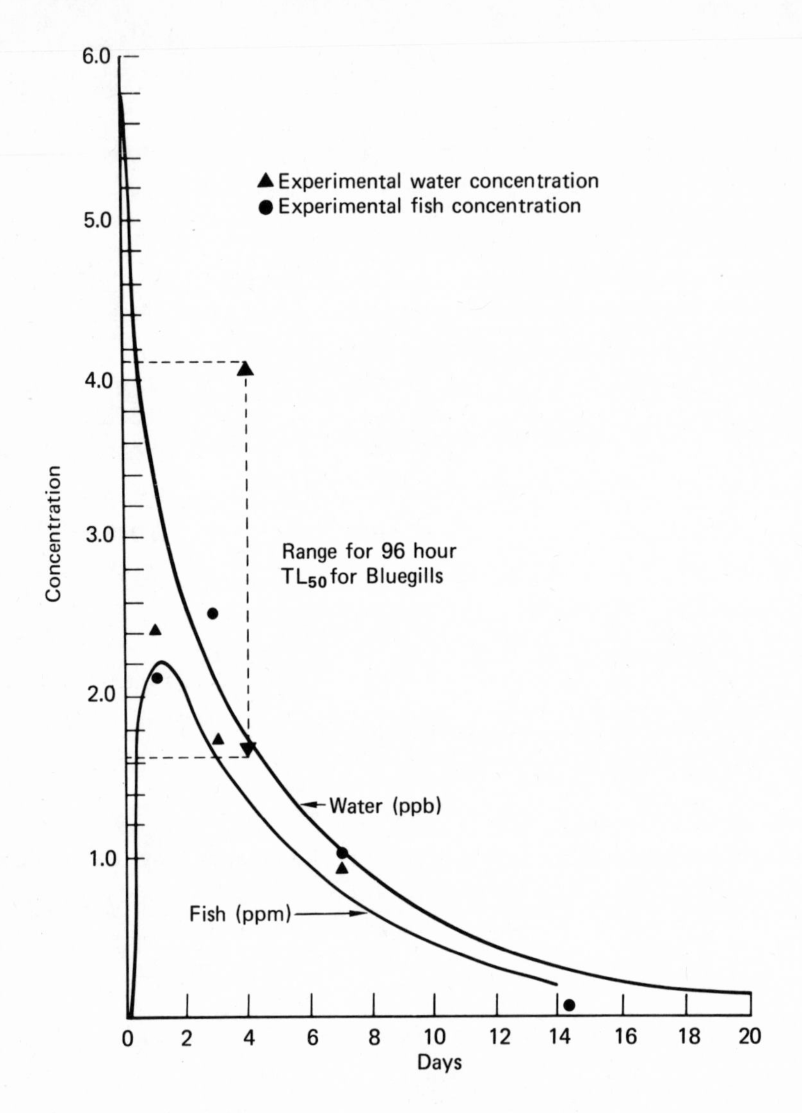

Figure 3. Computer simulation of the concentration of Chlorpyrifos in both fish and water. The experimental numbers are shown by (W) for water and (") for fish.

TABLE III

PARTITIONING OF CHLORPYRIFOS IN THE VARIOUS COMPARTMENTS OF THE FISH POND*

Time	% of Total in Various Compartments					
Time (Days)	Water	Soil and Plants	Fish	Air	Amount Metabolized	Amount Hydrolyzed
1	54	26	0.9	2.3	1.3	15
2	48	25	0.8	3.8	2.9	25
5	24	16	0.45	6.9	6.3	46
10	10.4	6.7	0.19	9.5	9.2	64
15	4.4	2.8		10.7	10.4	71
20	1.8	1.2		11.2	11	75
25	0.8	0.5		11.4	11	76

*These results were generated by the computer simulation of Equations 1 & 2 using the rate constants and pond parameters discussed in the report.

lower the water concentration below the critical level or c) formulate the insecticide to reduce the actual amount of insecticide that is in the water at any time. The effect of option (b) can be demonstrated by increasing the soil absorption rate constant by a factor of 5. The results of this experiment are shown in Figure 4. This causes the water concentration to be reduced to a level well below the 96-hour LC_{50} for bluegills.

Simulation 3

Assuming that the pond is devoid of living organisms and no absorption on soil occurs, what would the half-life of Chlorpyrifos be? This was examined by setting all the rate constants equal to zero except for hydrolysis and evaporation. The results are shown in Figure 5 where the half-life of the insecticide is ∿4 days. This supports the observation that long-term chronic effects would not be an important consideration in the use of this insecticide under these conditions.

Bioconcentration

The bioconcentration potential in fish is usually expressed as the ratio of the concentration of chemical in the fish to the concentration in water at steady state where C_w is constant. In the situation described in this report, a different steady state is reached due to the adsorption and release of chemical from the soil. An apparent bioconcentration factor may be estimated by the following procedure. The material balance equations for the insecticide in water, soil and fish are as follows:

$$\frac{dC_w}{dt} = -k_1 C_w/d - k_2 C_w - k_3 F C_w/V - k_5 S C_w/V + k_6 S C_S/V \quad (3)$$

$$\frac{dC_S}{dt} = k_5 C_w - k_6 C_S \quad (4)$$

$$\frac{dC_f}{dt} = k_3 C_w - k_4 C_f \quad (5)$$

Equations 3 and 4 may be solved simultaneously to express the water concentration as a function of time.

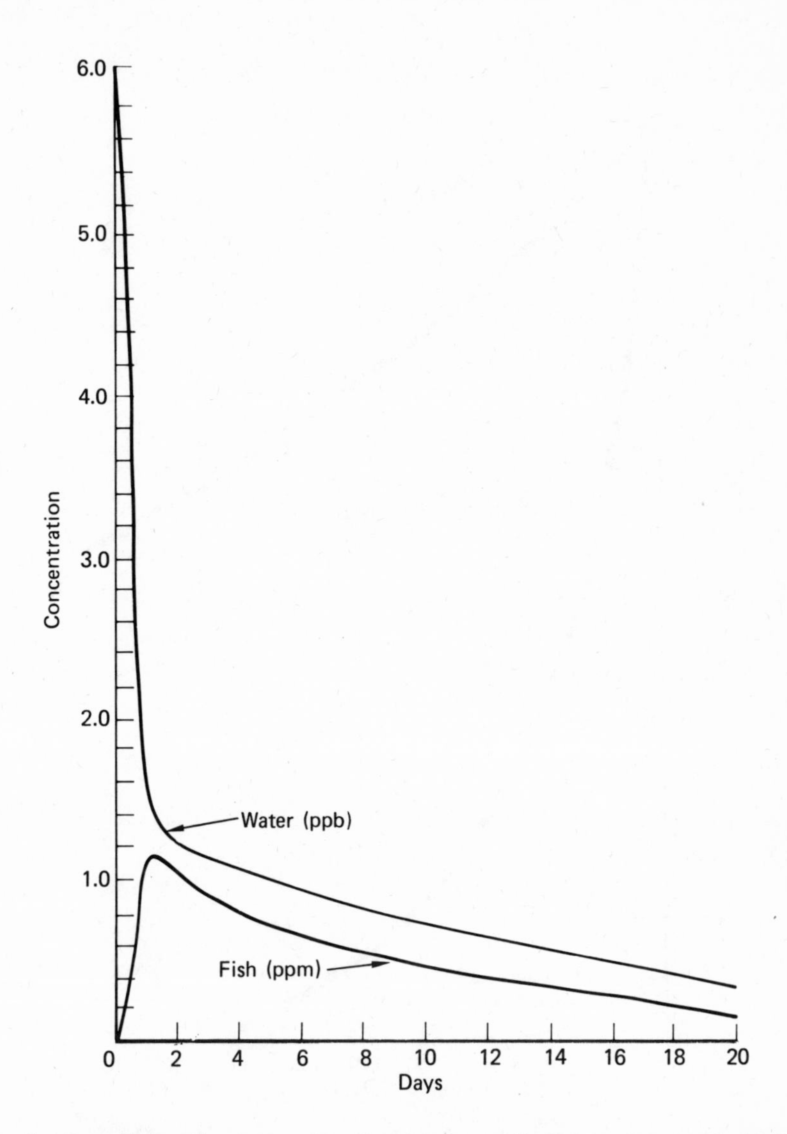

Figure 4. Results of computer simulation of the water and fish concentration when the soil absorption constant was increased by a factor of 5.

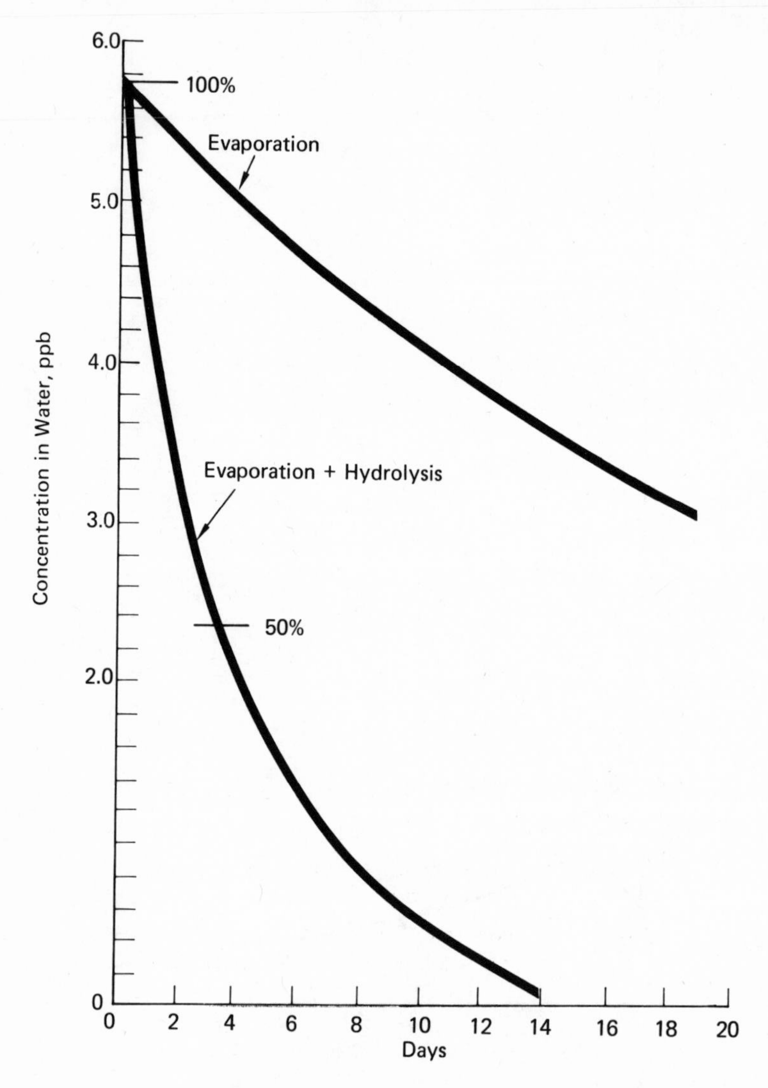

Figure 5. Computer simulation of disappearance of Chlorpyrifos subjected to hydrolysis and evaporation.

$$C_w = \frac{(k_6-\beta)C'_w\exp(-\alpha t)}{(\alpha-\beta)} - \frac{(k_6-\alpha)C'_w\exp(-\alpha t)}{(\alpha-\beta)} \qquad (6)$$

where $\alpha = [Q + [Q^2 - 4\ k_6(k_1/D + k_2 + k_3F/V)]^{1/2}]/2$

$\beta = [Q - [Q^2 - 4\ k_6(k_1/D + k_2 + k_3F/V]^{1/2}]/2$

$Q = k_1/d + k_2 + k_3F/V + k_5S/V + k_6$

C'_w = initial concentration of chemical in the water.

Substituting for C_w in equation 5 and integrating the following expression for the time course of chemical in the fish is given.

$$C_f = \frac{k_3(k_6-k_4)C_w'\exp(-k_4t)}{(\alpha-k_4)\quad(\beta-k_4)} + \frac{k_3(k_6-\alpha)C_w'\exp(-\alpha t)}{(k_4-\alpha)\quad(\beta-\alpha)} \qquad (7)$$

$$+ \frac{k_3(k_6-\beta)C_w'\ \exp(-\beta t)}{(k_4-\beta)\quad(\alpha-\beta)}$$

Since both equations 6 and 7 involve the combination of exponential terms, recall that $\alpha>\beta$ implies that the exponential term in α goes to zero before the term in β as time increases. In addition, the quantity β numerically approaches k_6, the rate constant for release of chemical from soil and plants and α numerically approaches the sum of the rate constants for all dissipating mechanisms. The ratio of the concentration of Chlorpyrifos in fish to the concentration in water is obtained by dividing equation 7 by equation 6.

$$\frac{C_f}{C_w} = \frac{k_3}{(k_4-\beta)} + \frac{k_3(\alpha-\beta)\exp(\beta-k_4)t}{(\alpha-k_4)\quad(k_6-\beta)} \qquad (8)$$

It is interesting to examine this ratio as time increases for the following special cases:

Case I $\beta > k_4$. This is the situation where the release from soil is faster than the elimination of metabolite from fish. Here the sign of the exponential term in equation 8 is positive with the result that C_f/C_w becomes increasingly large as time increases. Consequently steady state is never reached and a bioconcentration factor as described previously is not defined.

Case II $k_4 > \beta$. In this situation the sign of the exponential is negative so that the second term in equation 8 vanishes as time increases given $C_f/C_w = k_3/(k_4 - \beta)$. This is the bioconcentration factor at steady state. This value is larger than would be determined in a steady state bioconcentration test as described by Branson et al.[9] where β is zero.

Case III $k_6 = 0$. This is the situation if the material absorbed by the soil and plants is completely metabolized or stored and there is essentially no release of the parent material back into the water. In such a case α defined by equation 6 is now the sum of the rate constants characterizing the dissipating process and $\beta = 0$. The ration of C_f/C_w is given by:

$$\frac{C_f}{C_w} = \frac{k_3[\exp(\alpha - k_4)t - 1]}{(\alpha - k_4)} \tag{9}$$

As $t \to \infty$ $C_f/C_w \to k_3/(k_4 - \alpha)$ providing $k_4 > \alpha$. If $\alpha > k_4$, a situation that might arise if dissipating reactions were faster than the rate of metabolism, the C_f/C_w approaches infinity.

These examples indicate the extreme care that must be exercised in interpreting bioconcentration ratios from field observations. Without any prior knowledge of the initial conditions or the chemical and physical properties of the agent under study, it is impossible to decide whether a net steady state has been reached so that a bioconcentration factor can be defined.

It should be emphasized that while the bioconcentration factor for Chlorpyrifos is $\sim$700 in whole fish, the absolute amount that is present at any one time is significantly less when any of the normally occurring dissipating mechanisms are operating. The value of 700 is obtained from the ratio of the two rate

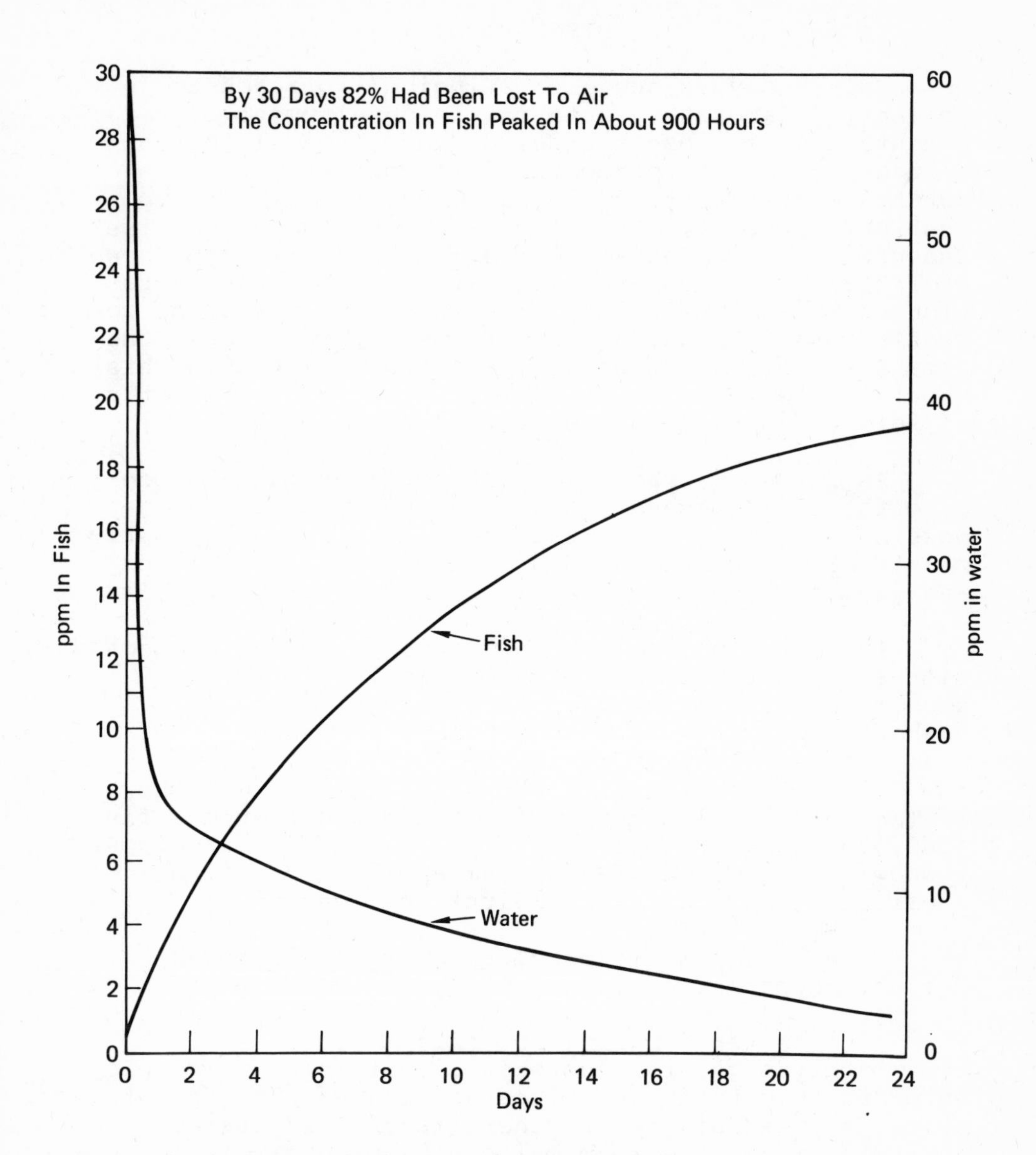

Figure 6. Computer simulation of water and fish concentration of a chemical with "DDT-like" properties.

constants k_3/k_4'. It is interesting to note that this value agrees with our previous estimate[8].

DISCUSSION

This analysis suggests a logical approach to collecting laboratory data in the development of a potentially new insecticide, or any chemical that has a high probability of coming into contact with an aquatic environment. We will illustrate with a hypothetical material, with properties that are "DDT-like". The properties of this hypothetical material are shown in Table IV. Once these are obtained a bioconcentration study as outlined by Branson et al.[9] should be conducted to determine rate constants characterizing the uptake and clearance of chemical from the fish. This should be followed by an investigation of the kinetics of soil absorption and desorption.

As far as we know there is not a standard procedure for making these soil measurements. It is possible, however, to obtain such data from an experiment similar to the one designed by Smith et al.[7] and analyzed by the procedures described by Blau et al.[1,3]

In order to obtain some knowledge of the biodegradation and the toxicity of the chemical to fish, a standard Biological Oxygen Demand (BOD) test and a 96-hour acute LC_{50} to fish should be obtained.

Hypothetical results for these results are shown in Table IV. Using these values and the previous fish pond, the curves shown in Figure 6 were simulated. Figure 7 shows the results of a second application of the chemical 60 days after the first application.

Several tentative conclusions may be reached for this hypothetical chemical.

1. The rapid buildup of this material in fish immediately suggests a chronic situation could develop. In order to satisfy ourselves as well as the regulatory people a great deal of further study is indicated to determine if such a chemical can be used safely.

2. The rapid loss of this agent to air indicates that air will become a major transport mechanism. It is interesting to note that with only a brief contact of the agent in water a long-term residual in fish occurs.

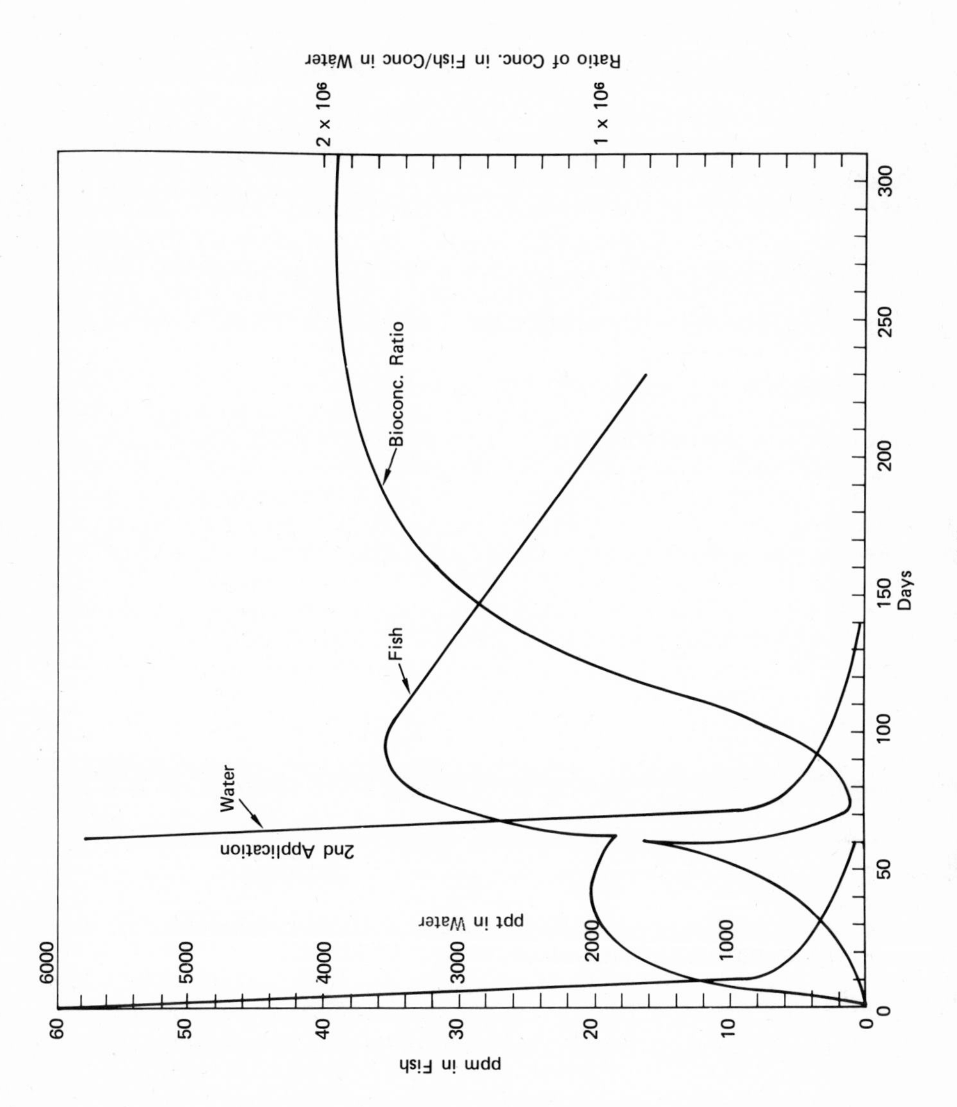

Figure 7. Computer simulation of water and fish concentration and bioconcentration ratio in fish as a result of two applications of a

TABLE IV

PROPERTIES OF A HYPOTHETICAL "DDT-LIKE" CHEMICAL

Property	Value
Solubility	1.2 ppb at 25°C[a]
Vapor pressure	1×10^{-7} mm Hg at 25°C[a]
Hydrolysis under natural water conditions	negligible[a]
Evaporation rate constant	9.34×10^{-1} cm/hour[a]
Partition Coefficient (octanol/water)	1×10^{6} [a]
Fish Studies	
Uptake rate constant	50 mℓ/g fish/hour[b]
Clearance[c]	2.5×10^{-4} hour^{-1} [b]
96-hour LC_{50}	1 ppb
Soil	
Uptake rate constant	3 mℓ/g soil/hour[b]
Release rate constant	0.052 hour^{-1} [a]
BOD	20 Day BOD ∿0[a]

[a] These are values for DDT.

[b] These are hypothetical values which probably are in the correct range for a DDT-like material.

[c] Hypothetical studies indicate that no metabolism took place in fish (the material was eliminated essentially unchanged).

3. The results of our prevous discussion on bioconcentration potential become immediately obvious when examining Figure 7. However, it is not possible to calculate this value from the previous equations because in our hypothetical example we assume no metabolism.

SUMMARY

When the various assumptions are recognized in translating the laboratory data concerning Chlorpyrifos to the fish ponds in Missouri, it is most satisfying to see the close agreement between calculated and observed concentrations.

It is our belief that relatively simple laboratory measurements can be translated to the real world and the predictions that are made have a high degree of credibility. Such models allow the visualization of the various flows that take place and are of great assistance in planning future experiments.

REFERENCES

1. Blau, G. E., and Neely, W. B., Ad. in Ecol. Research, 9, 133 (1975).

2. Macek, K. J., Walsh, D. F., Hogan, J. W., and Holz, D.D., Trans. Am. Fisheries Soc., 101, 420 (1972).

3. Blau, G. E., Neely, W. B., and Branson, D. R., Am. Inst. Chem. Eng., 21, 854 (1975).

4. Liss, P. S., and Slater, P. G., Nature, 247, 181 (1974).

5. Neely, W. B., Proc. of the 1976 National Conference on Control of Hazardous Chemical Spills. p. 197, April 1976.

6. Shaeffer, C. H., and Dupras, E. F., J. Econ. Entomol., 63, 701 (1970). These authors give several values for the rate constant. In conversation with D. Laskowski of the Agricultural Products Dept. of The Dow Chemical Company it was estimated that 3 days is a reasonable half-life for the hydrolysis of Chlorpyrifos in natural waters.

7. Smith, G. N., Watson, B. S., and Fisher, F. S., J. Econ. Entomol., 59, 1464 (1966).

8. Neely, W. B., Branson, D. R., Blau, G. E., Env. Sci. & Technol., 8, 1113 (1974).

9. Branson, D. R., Blau, G. E., Alexander, H. C. and Neely, W. B., Trans. Am. Fisheries Soc., 104, 785 (1975).

Section III

DEGRADATION OF PESTICIDES BY AQUATIC ORGANISMS

L. G. Hart, *Chairman*

AN INTRODUCTION TO PESTICIDE DEGRADATION BY AQUATIC ORGANISMS

L. G. Hart

I have been asked to chair the third and final section of this symposium on Pesticides in Aquatic Environments. This section will be concerned with the ways and means by which aquatic organisms degrade, or perhaps more correctly, biotransform or metabolize pesticides in their environment.

Pesticides, which may be insecticides, herbicides, nematocides, fungicides, rodenticides, etc., comprise a very large, heterogeneous group of chemicals. With the particular exception of the arsenicals, copper salts and fluorides, they are all organic chemicals. Like other foreign organic chemicals or xenobiotics, their pharmacologic or toxicologic actions are often modified by metabolism. However, the overall fate of a pesticide or other xenobiotic in an organism is dependent on a number of factors including absorption, distribution or storage, protein binding, excretion and metabolism. The role which each of these factors play is to a large extent affected by the physico-chemical properties of the chemical -- i.e., lipid solubility, degree of ionization at body pH, and presence or absence of reactive sites on the molecule. Since biological membranes are lipoid in nature, the more lipid soluble and less ionized molecules tend to traverse membranes more rapidly and completely in being absorbed and carried to sites of storage, metabolism or excretion. Of course, for metabolism to occur there has to be a reactive site(s) on the molecule.

Most pesticides, and other xenobiotics for that matter, are initially metabolized through one of four types of general reactions, namely oxidation, reduction, hydrolysis, or conjugation. The first three types have often been termed primary phase or phase I reactions. However, if reactive substituents such as hydroxyl or carboxyl groups are present on the parent compound then synthetic or conjugation reactions can occur initially. A classic example of this is the conjugation of benzoic acid with glycine to form hippuric acid. The products of the primary phase reactions are in general more polar than the parent compounds. However, the metabolite is not always less toxic or pharmacologically inactive, so these reactions can represent either detoxification, toxication, or activation. For example, oxidation of the thio-ether linkage of parathion results in the toxic product, paraoxon. Oxidation of aromatic or polycyclic hydrocarbons often produces reactive intermediates, epoxides or arene oxides, which are believed to be carcinogenic, mutagenic, or cytotoxic.

Perhaps the most important first phase reactions involving pesticides are (1) oxidative reactions, and (2) hydrolysis of esters, such as is associated with organophosphates and carbamates. With reference to the large class of organochlorine pesticides, reductive and GSH-mediated dechlorination reactions are also of some importance. The subsequent speakers will describe these where relevant, and I might add, they have been reviewed quite well recently by Dr. Matsumura in his excellent volume, Toxicology of Insecticides (1975), as well as in recent reports by Menzie (1974) and Khan et al. (1975).

Metabolic products of phase I reactions are then often susceptible to further metabolism by enzymes which catalyze synthetic or phase II reactions. Conjugating moieties include carbohydrates, amino acids, and other reactants, such as acetate and sulfate. The most important reactions in most species are glucuronide formation, glycine conjugation, mercapturic acid synthesis, methylation, acetylation, and ethereal sulfate synthesis. All of the four general types of reactions that I've mentioned have been reported to occur in marine and/or aquatic species. Studies in marine species in our laboratories have dealt primarily with oxidative or synthetic types of reactions.

Now, it is well known, that although species differences in type, degree, and duration of pharmacologic activity and/or toxicity may be accounted for by a number of factors, including those that I've mentioned, these differences are most often attributed to differences between species in the rates and sometimes the pathways of metabolism for xenobiotics.

To generalize -- Bend, James, and others in our laboratories (Bend et al., 1972; Bend et al., 1973; Pohl et al., 1973; Pohl et al., 1974; James et al., 1973; James et al., 1974); Adamson and coworkers (Adamson et al., 1965; Adamson, 1967; Adamson and Davies, 1973); Buhler and coworkers (Buhler and Rasmusson, 1968a; Buhler and Rasmusson, 1968b); and others have shown that most of the common marine species studied are deficient, relative to mammals, in their ability to metabolize or detoxify various xenobiotics. However, in making such a statement, I must say that the degree of deficiency is quite variable among marine species. For example, if we look at the P-450 dependent hepatic microsomal oxidative enzymes or MFOs, the little skate liver has comparable activity to that seen in rabbit liver for the hydroxylation of aniline with activity expressed as nmoles product formed/min/mg microsomal protein (Pohl et al., 1974). Yet, another elasmobranch, the dogfish shark, as well as certain teleosts and crustaceans have microsomal activity for this substrate that is an order of magnitude lower (Pohl et al., 1974). Adamson and coworkers (Adamson et al., 1965; Adamson, 1967) and Buhler and Rasmusson (Buhler and Rasmusson, 1968a) measured azo and nitro reductase activities in livers from a variety of marine and freshwater vertebrates, and found activities to be 5- to 10-fold or more lower than in rat liver. On the other hand, Bend, James, and Fouts (Bend and Fouts, 1973; James et al., 1974; James et al., 1976) have measured the activities for two enzyme systems which detoxify epoxides and arene oxides in the livers of several teleost and elasmobranch species, and found these activities, at least in teleosts, to be comparable to those in rabbit liver. The enzymes measured were epoxide hydrase, a microsomal enzyme and glutathione S-transferase, a cytosol enzyme.

Thus, the points I'd like to leave you with are (1) although differences in rates and pathways of metabolism usually may be the prime determinant in differences among species in pharmacologic or toxicologic response, one has to be careful in making such an assumption, and (2) many factors may modify metabolism, including such things as health, age, and diet of the organism, water temperature, and the presence of other xenobiotics.

REFERENCES

Adamson, R. H. 1967. Fed. Proc. 26: 1047-1055.

Adamson, R. H., R. L. Dixon, F. L. Francis, and D. P. Rall. 1965. Proc. Nat. Acad. Sci. 54: 1386-1391.

Adamson, R. H. and D. S. Davies. 1973. Comparative aspects of absorption, distribution, metabolism and excretion of drugs. In Comparative Pharmacology: Volume II, Section 85, International Encyclopedia of Pharmacology and Therapeutics, (M. J. Michelson, ed.). Pergamon Press, Oxford and New York, pp. 851-911.

Bend, J. R. and J. R. Fouts. 1973. Bull. Mt. Desert Island Biol. Lab. 13: 4-8.

Bend, J. R., R. J. Pohl, and J. R. Fouts. 1972. Bull. Mt. Desert Island Biol. Lab. 12: 9-12.

Bend, J. R., R. J. Pohl, and J. R. Fouts. 1973. Bull. Mt. Desert Island Biol. Lab. 13: 9-13.

Buhler, D. R. and M. E. Rasmusson. 1968a. Arch. Biochim. Biophys. 25: 223-239.

Buhler, D. R. and M. E. Rasmusson. 1968b. Comp. Biochem. Physiol. 25: 223.

James, M. O., J. R. Bend, and J. R. Fouts. 1973. Bull. Mt. Desert Island Biol. Lab. 13: 59-62.

James, M. O., J. R. Fouts, and J. R. Bend. 1974. Bull. Mt. Desert Island Biol. Lab. 14: 41-46.

James, M. O., E. W. Van Stee, and J. R. Bend. 1976. Fed. Proc. 35: 244.

Khan, M. A. Q., M. L. Gassman, and S. H. Ashrafi. 1975. Degradation of pesticides in biota. In Environmental Dynamics of Pesticides. (R. Haque and V. H. Freed, eds.) Plenum Press, New York and London, p. 378.

Matsumura, F. 1975. Toxicology of Insecticides. Plenum Press, New York and London, pp. 165-251, 325-354.

Menzie, C. M. 1974. Metabolism of Pesticides. An Update. U. S. Dept. Int., Fish and Wildlife Service, Special Scientific Report, Wildlife No. 184, Washington, D. C., p. 485.

Pohl, R. J., J. R. Bend, T. R. Devereux, and J. R. Fouts. 1973. Bull. Mt. Desert Island Biol. Lab. 13: 94-98.

Pohl, R. J., J. R. Bend, A. M. Guarino, and J. R. Fouts. 1974. Drug Metab. Disp. 2: 545-555.

XENOBIOTIC METABOLIZING ENZYMES IN MARINE FISH

M. O. James, J. R. Fouts, and J. R. Bend

In recent years, the interest of scientists from several disciplines has focused on the fate of the many xenobiotics, including pesticides, which are introduced into our environment. Many of these foreign chemicals eventually enter the oceans and can be ingested by marine species. Using *in vivo* and *in vitro* techniques, we have studied xenobiotic metabolism in a number of marine fish and crustacea. The pathways studied *in vitro* include oxidative metabolism by the cytochrome P-450-dependent mixed-function oxidases and metabolism of some products of oxidation, namely epoxides and arene oxides. A large number of pesticide molecules undergo metabolism by these reactions, which have been extensively studied in mammalian species. We also followed the uptake, distribution, and metabolism of a polychlorinated biphenyl, 2,4,5,2',5'-pentachlorobiphenyl, and of the herbicides, 2,4-dichlorophenoxyacetic acid and 2,4,5-trichlorophenoxyacetic acid, after administration to selected marine species.

MIXED-FUNCTION OXIDASE PATHWAYS

Several species of crustacea and elasmobranch and teleost fish were studied in marine laboratories in Maine and Florida. The animals were caught locally and maintained in tanks with fresh circulating sea water or in cages in tidal sea water until used. Immediately after sacrifice, livers were removed from the fish species and hepatopancreata from the crustacea, and the organs were placed in an ice-cold solution of 1.15% KCl/0.02 M HEPES, pH 7.4. Washed microsomes were prepared as described by Pohl et al. (1974), and protein concentrations in microsomes and cytosol fractions were determined by the method of Lowry et al (1951). Mixed-function oxidase activities toward benzphetamine, benzo(a)pyrene, and 7-ethoxycoumarin were measured by standard techniques (see Pohl et al., 1974). In three Florida species, the sheepshead, the drum, and the stingray, we found that these three mixed-function oxidase activities increased up to 35 to 40°C then declined sharply, whereas in two Maine species, the little skate and the dogfish shark, activities fell off above 30°C. Thus, we routinely assayed mixed-function oxidase activities at 35°C in Florida and 30°C in Maine species. Other variables in the assays, such as incubation time, reaction pH, concentrations of substrates, and cofactors, and amounts of protein used, were those which gave maximum activity in the little skate in Maine and the stingray and sheepshead in Florida. We found that, except for temperature of incubation and reaction pH, the conditions under which maximum activity was measured were similar for most species. Thus, all Maine species were assayed under the conditions which were optimum for the little skate; Florida elasmobranchs were assayed under conditions which were optimum for the stingray; and Florida teleosts were assayed under conditions which were established for the sheepshead. Cytochrome P-450 contents of microsomes were measured by the dithionite difference method, as described by Bend et al. (1972), using a Shimadzu spectrophotometer.

Tables 1 and 2 show our results for NADPH-supported mixed-function oxidase activities, and Table 3 shows the cytochrome P-450 contents of hepatic microsomes from several Maine and Florida species. In general, the teleost species had higher activities of benzpyrene hydroxylase than elasmobranchs, whereas the ranges of activities of 7-ethoxycoumarin O-deethylase and benzphetamine N-demethylase were the same for teleosts and elasmobranchs. Crustaceans had very low, usually undetectable, activities toward all three model substrates when NADPH was used as the cofactor. However, the cytochrome P-450 content of hepatopancreas microsomes was high, often higher than cytochrome P-450 content of fish livers, and hepatopancreas microsomes exhibit typical binding spectra with benzphetamine (type 1) and aniline (type 2) (Elmamlouk et al., 1974).

TABLE 1

XENOBIOTIC METABOLIZING ENZYMES IN HEPATIC MICROSOMES FROM SOME FLORIDA MARINE SPECIES

SPECIES	Activity[a]		
	Benzpyrene hydroxylase F.U.	7-Ethoxycoumarin O-deethylase (nmol)	Benzphetamine N-demethylase (nmol)
TELEOSTS			
Sheepshead			
Archosargus probatocephalus	1.38 ± 0.47 (9)	0.05 ± 0.03 (9)	1.10 ± 0.42 (9)
Drum			
Pogonias cromis	1.62 ± 1.91 (11)	0.06	0.45 ± 0.09 (5)
Jack crevalle			
Caranx hippos	2.31		
Mangrove snapper			
Lutjanus griseus	6.64 ± 0.48 (3)[b]	0.16 ± 0.02 (3)[b]	0.88, 2.50 (2)[b]
ELASMOBRANCHS			
Atlantic stingray			
Dasyatis sabina	0.86 ± 0.36 (27)	0.05 ± 0.02 (14)	0.98 ± 0.75 (16)
Bluntnose ray			
Dasyatis sayi	0.17 ± 0.05 (3)	0.01, N.D.,N.D.(3)	0.23 ± 0.11 (3)
CRUSTACEA			
Spiny Lobster			
Panulirus argus	N.D. to 0.04 (15)	N.D. (6)	N.D. (4)
Blue crab			
Callinectes sapillus	N.D. to 0.01 (4)		

[a]Expressed as units indicated per min per mg protein, mean ± S.D. (n). Individual values are given where less than three points were obtained. N.D. means not detected.
[b]Pools of livers from up to 15 fish were needed to obtain sufficient microsomal protein for assay.

TABLE 2

XENOBIOTIC METABOLIZING ENZYMES IN HEPATIC MICROSOMES OF SOME MAINE MARINE SPECIES

SPECIES	Activity[a]		
	Benzpyrene hydroxylase F.U.	7-Ethoxycoumarin O-deethylase (nmol)	Benzphetamine N-demethylase (nmol)
TELEOSTS			
Eel *Anguilla rostrata*	0.21[b]	0.89[b]	0.44[b]
Mummichog *Fundulus heteroclitus*	4.21 ± 2.10 (3)[b]	0.47 ± 0.30 (3)[b]	1.13 ± 0.76(3)[b]
Winter flounder *Psuedopleuronectes americanus*	2.54 ± 1.67 (7)	0.32 ± 0.25 (6)	0.59 ± 0.13(3)
King of Norway *Hemitripterus americanus*	0.004, 0.02	0.06 ± 0.08 (3)	0.16 ± 0.04(3)
ELASMOBRANCHS			
Dogfish shark *Squalus acanthias*	0.07 ± 0.02 (3)	0.08 ± 0.02 (3)	0.15 ± 0.05(3)
Little skate *Raja erinacea*	0.17 ± 0.10 (10)	0.32 ± 0.14 (11)	1.07 ± 0.19(9)
Thorny skate *Raja radiata*	0.12 ± 0.02 (3)	0.12 ± 0.04 (3)	0.45 ± 0.14(3)
Large skate *Raja ocellata*	0.30 ± 0.09 (3)	0.47 ± 0.08 (3)	1.49 ± 0.41(3)
CRUSTACEA			
Lobster *Homarus americanus*	N.D. to 0.02 (6)	N.D. (6)	N.D. (6)

[a]Expressed as units indicated per min per mg protein, mean ± S.D. (n). Individual values are given where less than three points were obtained. N.D. means not detected.
[b]Pools of livers from up to 15 fish were needed to obtain sufficient microsomal protein for assay.

TABLE 3

CYTOCHROME P-450 CONTENT IN HEPATIC MICROSOMES OF SEVERAL MARINE SPECIES

SPECIES	Cytochrome P-450 Concentration (nmoles/mg microsomal protein)
FLORIDA	
TELEOSTS	
Sheepshead	0.28 ± 0.09 (9)[a]
Drum	0.14 ± 0.03 (4)
Mangrove snapper	0.25 ± 0.05 (3)[b]
ELASMOBRANCHS	
Atlantic stingray	0.43 ± 0.07 (8)
Bluntnose ray	0.32 ± 0.12 (3)
CRUSTACEA	
Spiny lobster	0.88 ± 0.52 (15)
Blue crab	N.D. to 0.29 (4)[c]
MAINE	
TELEOSTS	
Winter flounder	0.17 ± 0.01 (3)
ELASMOBRANCHS	
Dogfish shark	0.23, 0.29
Little skate	0.28 ± 0.05 (4)
Large skate	0.36, 0.41

[a]Mean ± S.D. (n). Individual values are given where less than 3 animals were assayed.

[b]Pools of livers from up to 15 fish were needed to obtain sufficient microsomal protein.

[c]N.D. means not detected.

Hepatopancreas microsomes and cytosol fractions were found to inhibit mixed-function oxidase activities (especially benzpyrene hydroxylase) of sheepshead and stingray microsomes (Table 4). Possibly an inhibitory factor is released during the disruption of the hepatopancreas cells in the preparation of microsomes and binds to some component of the NADPH-dependent cytochrome P-450 mixed-function oxidase system. Dialysis of microsomes or cytosol fraction did not remove the inhibitory factor, but boiling for 10 minutes did destroy most of the inhibitory action of cytosol fraction. The nature of this inhibitory factor is not known, but we are currently attempting to characterize it.

Pretreatment of sheepshead with 3-methylcholanthrene (3-MC), at a dose of 20 mg/kg intraperitoneally, caused a tenfold increase in hepatic microsomal benzpyrene hydroxylase activity (see Table 5) and a roughly fivefold increase in 7-ethoxycoumarin O-deethylase activity but did not affect the cytochrome P-450 contents or benzphetamine N-demethylase activity of the treated animals with respect to the controls. By contrast, up to three doses of 3-MC (20 mg/kg, i.p.) to the stingray, another species studied in Florida, did not affect benzpyrene hydroxylase or 7-ethoxycoumarin O-deethylase activities, and again cytochrome P-450 contents and benzphetamine N-demethylase activities were unchanged. 3-MC was poorly absorbed by these fish since we observed 3-MC in the peritoneal cavity of each stingray at the time of sacrifice throughout the course of the experiment, which extended to 30 days after the last dose (42 days after the first dose).

Administration of 2,3,7,8-tetrachlorodibenzo-*p*-dioxin (TCDD), a very potent inducer of arylhydrocarbon hydroxylase activity in mammalian species, induced benzpyrene hydroxylase activity in the flounder and the little skate, two species which were studied in Maine. Administration of 3-MC to the flounder and the little skate has yielded a variable response (Pohl et al., 1974). Sometimes hepatic activities of 7-ethoxycoumarin O-deethylase and benzpyrene hydroxylase were higher in treated animals than in control animals, whereas at other times the 3-MC treatment had no effect on hepatic activities. Poland et al. (1974) have shown that TCDD induces aryl hydrocarbon hydroxylase activity in so-called non-responding mice (strains of mice that do not exhibit increased aryl hydrocarbon hydroxylase activity after 3-MC treatment). The variable response to 3-MC but uniform response to TCDD which we observed

TABLE 4

INHIBITION OF FISH ACTIVITY BY HEPATOPANCREAS PREPARATIONS FROM SPINY LOBSTER

mg Hepatopancreas protein added	Benzpyrene hydroxylase activity F.U./min/mg fish microsomes[a]	
Hepatopancreas microsomes	Sheepshead (1)	Sheepshead (2)
0	1.83	2.79
1.02	0.75	0.82
1.20	0.56	0.44
Hepatopancreas cytosol	Sheepshead (3)	Sheepshead (4)
0	1.78	2.63
1.20	0.07	0.08
2.40	0.05	0.06
6.00	0.01	0.02
Hepatopancreas microsomes	Ray (1)	Ray (2)
0	1.22	0.88
1.16	0.29	0.23
2.92	0.02	0.03

[a]Sheepshead or ray hepatic microsomes were added to different amounts of hepatopancreatic microsomes or cytosol fractions and the mixtures used in the benzpyrene hydroxylase assay. Activities were calculated based on the amount of sheepshead or ray microsomes present.

TABLE 5

EFFECT OF PRETREATMENT WITH INDUCING AGENTS

SPECIES	TREATMENT (number)	Cytochrome P-450[a]	Benzpyrene hydroxylase[b]	7-Ethoxycoumarin O-deethylase[c]
Sheepshead	Corn oil (9)[d]	0.28 ± 0.09	1.41 ± 0.49	0.051 ± 0.028
	3-Methylcholanthrene (6)[e]	0.44 ± 0.17	15.21 ± 8.19	0.227 ± 0.080
Stingray	Corn oil (5)[d]	0.43 ± 0.07	0.60 ± 0.24	0.040 ± 0.007
	3-Methylcholanthrene (5)[f]	0.40 ± 0.12	0.75 ± 0.67	0.050 ± 0.028
Little Skate	Corn oil:acetone (6)[d]	0.24 ± 0.02	1.22 ± 0.17	0.360 ± 0.058
	TCDD (5)[g]	0.25 ± 0.03	18.4 ± 3.00	0.563 ± 0.096
Flounder	Corn oil:acetone (6)[d]		36.7 ± 12.8	0.199 ± 0.068
	TCDD (4)[h]		59.2 ± 4.6	0.418 ± 0.070
Little Skate	Corn oil (3)[d]	0.16 ± 0.03	0.61 ± 0.19	
	Dibenzanthracene (3)[i]	0.15 ± 0.04	5.04 ± 2.26	

[a]Nmol/mg microsomal protein.
[b]F.U./min/mg protein.
[c]Nmol/min/mg protein.
[d]Controls received the solvent vehicle by the same route and at the same time as the treated animals.
[e]Two doses of 3-MC (20 mg/kg, i.p.) on days 1 and 3, animals sacrificed on day 6
[f]Two doses of 3-MC (20 mg/kg, i.p.) of days 1 and 4, animals sacrificed on day 6.
[g]Two doses of TCDD (4.5 μg/kg, i.p.) on days 1 and 3, animals sacrificed on day 7.
[h]Two oral doses of TCDD (4.5 μg/kg) on days 1 and 3, animals sacrificed on day 8.
[i]Two doses of dibenzanthracene (10 mg/kg, i.p.) on days 1 and 3, animals sacrificed on day 8.

in the flounder and little skate may be a reflection of the heterogeneity of the wild population with respect to this response to chemicals or to induction in the field as a result of exposure to environmental contaminants. Dibenzanthracene, a less toxic chemical than TCDD but one which has similar effects on aryl hydrocarbon hydroxylase activity in mammals, also induced benzpyrene hydroxylase and 7-ethoxycoumarin O-deethylase activities in the little skate without affecting cytochrome P-450 content.

Mixed-function oxidase activities towards benzpyrene, benzphetamine, and 7-ethoxycoumarin were also measured using cumene hydroperoxide or sodium periodate in place of an NADPH generating system, since these oxidizing agents have been shown to support mixed-function oxidase activities in the presence of cytochrome P-450 (Hrycay et al., 1975). When these agents are used, other components which are necessary for activity with NADPH, such as lipid and NADPH-cytochrome P-450 reductase, are not needed (Coon et al., 1975). In support of this work, we found that in using cumene hydroperoxide or sodium periodate with hepatopancreas microsomes we were able to measure oxidative metabolism of three model substrates.

A comparison of the rates of formation of oxidized products using NADPH, cumene hydroperoxide, or sodium periodate with microsomes from the spiny lobster, two species of stingrays, the mangrove snapper and the sheepshead, are shown in Table 6. In all species studied, this activity was concentrated in the microsomes, as was cytochrome P-450, and could be abolished by boiling the microsomes prior to assay. In no case was product formed in the absence of tissue. Hepatopancreas cytosol fraction did not inhibit cumene hydroperoxide or sodium periodate supported activity of sheepshead microsomes, supporting the idea that the inhibitor does not affect the initial binding of the substrate to cytochrome P-450.

The ratios of activities for the three cofactors were different for each substrate. For most fish species, formation of fluorescent phenolic metabolites from benzpyrene was supported best with NADPH and least well by cumene hydroperoxide. In six experiments with pooled sheepshead hepatic microsomes and cumene hydroperoxide, no fluorescent metabolites of benzpyrene were detected. However, cumene hydroperoxide did support benzphetamine N-demethylase and 7-ethoxycoumarin

TABLE 6

MIXED-FUNCTION OXIDASE ACTIVITY SUPPORTED BY CHEMICALS OTHER THAN NADPH

SPECIES	AGENT[a]	Benzpyrene hydroxylase[b]	Benzphetamine N-demethylase[c]	7-Ethoxycoumarin O-deethylase[c]
SHEEPSHEAD (6 pools)[d]	NADPH	1.62 ± 0.49	0.58 ± 0.30	0.08 ± 0.03
	Cum-OOH	< 0.01	7.9 ± 4.4	0.06 ± 0.03
	$NaIO_4$	0.36 ± 0.10	113 ± 42	1.45 ± 1.16
MANGROVE SNAPPER (3 pools)[e]	NADPH	6.3 ± 0.5	2.5, 0.88	0.17 ± 0.01
	Cum-OOH	0.02, 0.04[f]	6.1, 6.6[f]	0.07 ± 0.06
	$NaIO_4$	0.44 ± 0.05	216, 245[f]	1.65 ± 0.53
BLUNTNOSE RAY (n = 3)	NADPH	0.21, 0.13[f]	0.34, 0.12[f]	0.01, N.D.[f]
	Cum-OOH	0.04 ± 0.02	3.91 ± 0.44	0.01, 0.03[f]
	$NaIO_4$	0.33, 0.36[f]	152, 139[f]	0.13[g]
STINGRAY (5 pools)[d]	NADPH	0.75 ± 0.06	0.38 ± 0.12	0.06 ± 0.02
	Cum-OOH	0.04, N.D.[f]	8.9 ± 6.7[f]	0.16, 0.07[f]
	$NaIO_4$	0.47 ± 0.37	100 ± 50	0.90 ± 0.18
SPINY LOBSTER (5 pools)[d]	NADPH	0.00 to 0.04	N.D.	N.D.
	Cum-OOH	0.06 ± 0.02	14.9 ± 7.5	0.32 ± 0.12
	$NaIO_4$	0.36 ± 0.07	189 ± 47	0.58 ± 0.28

[a]Chemical added to initiate oxidation of substrate by microsomes. NADPH = nicotinamide adenine-dinucleotide phosphate, Cum-OOH = cumene hydroperoxide, $NaIO_4$ = sodium periodate.
[b]Activity is expressed in F.U./min/mg protein, mean ± S.D.
[c]Activity is expressed in nmol/min/mg protein, mean ± S.D.
[d]Up to four animals were pooled to prepare microsomes.
[e]Twelve to sixteen animals were pooled to prepare microsomes.
[f]Assays were performed on only two pools of microsomes.
[g]Assays were performed on only one pool of microsomes.

O-deethylase activities with sheepshead hepatic microsomes. This apparent lack of benzpyrene hydroxylase activity may be due to the oxidation of phenolic metabolites.

It was especially interesting to find cumene hydroperoxide or sodium periodate-supported mixed-function oxidase activity in hepatopancreas microsomes since we know that xenobiotics are slowly metabolized *in vivo* by crabs and lobsters even though NADPH-dependent activity is undetectable for most substrates.

EPOXIDE METABOLIZING ACTIVITIES

The work described so far shows that hepatic microsomes from the majority of fish species studied are able to oxidize xenobiotics by the NADPH-dependent mixed-function oxidase enzyme complex and that hepatic or hepatopancreatic microsomes from fish or crustacea can support oxidative metabolism in the presence of cumene hydroperoxide or sodium periodate, although the *in vivo* significance of this *in vitro* activity is presently unknown. Thus, most species studied are capable of forming potentially toxic epoxide or arene oxide metabolites. Epoxides can be further metabolized by two major pathways, microsomal epoxide hydrase and cytosol fraction glutathione S-transferase. These are usually considered detoxication pathways, although some evidence suggests that epoxide hydrase can act on certain polycyclic arene oxides to produce carcinogenic metabolites (Wislocki et al., 1976). The glutathione S-transferases are a family of enzymes with overlapping substrate specificities and a large number of substrates, both endogenous and exogenous. (For a review of the function and properties of the glutathione S-transferases, see L. F. Chasseud, 1976.) The glutathione S-transferase family of enzymes is ubiquitous among species so far studied from all classes and probably constitutes an important mechanism for detoxication of a number of xenobiotics.

We measured epoxide hydrase and glutathione S-transferase activities *in vitro* using styrene oxide as substrate. The assay procedures for styrene oxide are essentially those described in James et al. (1976). Epoxide hydrase activity was measured at 30°C in Maine and at 35°C in Florida species to correspond to the temperatures at which mixed-function oxidase activities were measured, although for the two fish species studied in each location (flounder and skate in Maine, sheepshead and stingray in Florida), the temperature optima using styrene oxide as substrate were higher than this (40 to 45° for both Maine species and the stingray in Florida, 45 to 50° for the sheepshead in Florida). Hepatopancreatic microsomes from the spiny lobster had a broad temperature optimum for epoxide hydrase activity between 35 and 45°C. Epoxide hydrase activity toward styrene oxide in skate, ray, and sheepshead hepatic microsomes was linear with time for up to 45 minutes and the incubation time was usually 15 minutes. Glutathione S-transferase activity toward styrene oxide was linear with time for ten to fifteen minutes in the stingray or sheepshead cytosol fractions. Incubation times used were ten minutes for the Maine species and five minutes for the Florida species. Substrate concentrations for epoxide hydrase and glutathione S-transferase assays were 1 mM for styrene oxide. Glutathione S-transferase activity was assayed in the presence of 10 mM glutathione. Maximum in vitro epoxide hydrase activity was measured at pH 8.7 in sheepshead, stingray, and skate, although activity did not vary greatly with pH in the range pH 8 to pH 9.5. The pH-activity profile of glutathione S-transferase activities toward styrene oxide showed two maxima in the range pH 6.5 to 8.3, one at pH 6.8 to 7.0, and one at pH 7.6 to 7.8, though again the variation of activity in this range was small. Above pH 8, nonenzymatic reaction of epoxide with glutathione proceeded fairly rapidly. Glutathione S-transferase activity was assayed, pH 7.6.

A species comparison of hepatic epoxide hydrase and glutathione S-transferase activities is given in Table 7 and 8. Both activities could be easily measured in all species tested. In most species, specific activities of glutathione S-transferase were higher than specific activities of epoxide hydrase, as is the case with most mammalian species. However, epoxide hydrase activities in hepatopancreas microsomes from crustacea were usually higher than glutathione S-transferase activities in the hepatopancreas cytosol fraction. It was somewhat surprising to find such high in vitro activities of epoxide hydrase in the species which had undetectable NADPH-dependent mono-oxygenase activities. Perhaps some other mechanism exists in crustacea for the formation of epoxides. The high epoxide hydrase activity of crustacean hepatopancreas suggests that endogenous compounds such as hormones

TABLE 7

HEPATIC EPOXIDE HYDRASE AND GLUTATHIONE S-TRANSFERASE ACTIVITIES IN MARINE TELEOSTS WITH STYRENE OXIDE AS SUBSTRATE

SPECIES	Epoxide Hydrase Activity[a]	Glutathione S-Transferase Activity[b]
FLORIDA FISH		
Sheepshead	6.1 ± 2.6 (16)[c]	25.5 ± 7.7 (16)
Drum	4.7 ± 2.0 (13)	16.1 ± 5.2 (13)
Mangrove snapper	2.5, 1.6[d]	4.2 ± 0.8 (4)[d]
Black sea bass *Gentropristes striatus*	0.6	2.1
Southern flounder *Paralichthys lethostigma*	2.1	6.6
MAINE FISH		
Winter flounder	2.0 ± 1.4 (4)	4.4 ± 0.9 (4)
King of Norway	1.8 ± 0.4 (3)	2.6 ± 0.5 (3)

[a]Nmoles styrene glycol formed/min/mg microsomal protein.

[b]Nmoles S-(2-hydroxy-1-phenylethyl)glutathione formed/min/mg 176,000xg supernatant fraction protein.

[c]Data expressed as mean ± S.D. (n). Individual values are given where less than 3 fish were assayed.

[d]Pools of liver from up to 15 fish were required to obtain sufficient microsomal protein for assay.

TABLE 8

HEPATIC EPOXIDE HYDRASE AND GLUTATHIONE S-TRANSFERASE ACTIVITIES IN MARINE ELASMOBRANCHS AND CRUSTACEA WITH STYRENE OXIDE AS SUBSTRATE

SPECIES	Epoxide Hydrase Activity[a]	Glutathione S-Transferase Activity[b]
ELASMOBRANCHS (Florida)		
Atlantic stingray	7.6 ± 0.7 (8)[c]	5.1 ± 1.6 (8)
Bluntnose ray	2.4, 4.7	4.8
(Maine)		
Little skate	0.4 ± 0.2 (17)	2.4 ± 0.5 (17)
Large skate	1.8 ± 0.2 (5)	2.6 ± 0.3 (3)
Thorny skate		7.1 ± 2.8 (4)
Dogfish shark	7.6 ± 2.4 (4)	9.3
CRUSTACEA (Florida)		
Blue crab	7.5 ± 3.2 (3)	0.4 ± 0.3 (3)
Spiny lobster	23.4 ± 4.4 (8)	1.1 ± 0.3 (8)
(Maine)		
Rock crab	15.1 ± 3.1 (4)	0.3 ± 0.2 (4)
Lobster	21.9 ± 2.4 (4)	1.3 ± 0.6 (4)

[a]Nmoles styrene glycol formed/min/mg microsomal protein.

[b]Nmoles S-(2-hydroxy-1-phenylethyl)glutathione formed/min/mg 176,000xg supernatant.

[c]Data expressed as mean ± S.D. (n). Individual values are given when less than 3 animals were assayed.

are natural substrates for epoxide hydrase and also raises the possibility that the dihydrodiol products are physiologically important in crustacea.

3-Methylcholanthrene or 2,3,7,8-tetrachlorodibenzo-p-dioxin pretreatment of sheepshead, ray, skate, or flounder had no effect on the measured activity of epoxide hydrase or glutathione S-transferase towards styrene oxide or benzo(a)pyrene 4,5-oxide. We are currently testing the effect of two other chemical agents, phenylbutazon and Arochlor 1254, on epoxide metabolizing activities.

TISSUE DISTRIBUTION OF POLYCHLORINATED BIPHENYLS

As a class of chemicals, the polychlorinated biphenyls (PCBs), are major environmental contaminants, due to their widespread use and high chemical stability. We have studied the uptake, distribution, and metabolism in lobster and dogfish shark of one of the components of Arochlor 1254, namely 2,4,5,2',5'-pentachlorobiphenyl (2,4,5,2',5'-PCB), using the 14-C labeled compound, which was obtained from Mallinckrodt Nuclear. Details of the dosage regimens used and the methods of analysis are given in Bend et al. (1976). We found that most (about 90%) of the dose of 14-C 2,4,5,2',5'-PCB was taken up from hemolymph by the hepatopancreas of the lobster (Homarus americanus). Similarly, about 90% of an i.v. dose of 2,4,5,2'.5'-PCB to the dogfish shark (Squalus acanthias) was taken up from blood into liver. The chemical was eliminated very slowly by either species. In the lobster, the sex organs also took up injected 2,4,5,2',5'-PCB from hemolymph. The estimated half-life of the PCB isomer in lobster hepatopancreas and female egg masses was about 40 days, and in the male gonads about 7 days. Dogfish shark could not be maintained in captivity for longer than 12 days and at this time the 14-C PCB content of liver was the same as was found at 6 hours after dosing. The decline in radioactivity in the dogfish shark blood after intravenous injection of 14-C 2,4,5,2',5'-PCB was very rapid and related to the rapid uptake of the 14-C-PCB into liver. Dogfish shark liver and lobster hepatopancreas both have high lipid content, and the accumulation of the lipid soluble 2,4,5,2',5'-PCB into these organs was not surprising. Almost all (greater than 90%) of the 14-C present in liver or hepatopancreas at every time point studied was present as the parent compound. The slow rate of metabolism of this PCB isomer in dogfish shark and lobster, compared with rats, agrees with the low *in vitro* mono-oxygenase activities found in liver of dogfish shark and hepatopancreas of lobster (Table 2). Trace quantities of water soluble metabolites were found in the green gland and feces, but were not characterized.

METABOLISM OF CARBOXYLIC ACIDS

Another class of chemicals in widespread use which occur as pollutants in the marine environment is the carboxylic acid herbicides, for example, 2,4-dichlorophenoxyacetic acid and 2,4,5-trichlorophenoxyacetic acid. These two compounds are excreted largely unchanged by most mammals, although small amounts (less than 5% of the dose) are excreted by rats as glycine and taurine conjugates (Grunow and Bohme, 1974). This is in contrast to the structurally related phenylacetic acid and 4-chloro- and 4-nitrophenylacetic acids, which are all extensively (greater than 90%) conjugated with glycine by rats prior to excretion in urine (James et al., 1972a). We wanted to determine whether phenylacetic acid was conjugated with an amino acid or amino acids prior to excretion by fish species, since the metabolism of this compound is subject to wide species variation (James et al., 1972b). Subsequently, we studied the metabolism of the structurally related phenoxyacetic acid herbicides. Some earlier studies with 2,4-dichlorophenoxyacetic acid (2,4-D) and 2,4-D-butoxyethanol ester were carried out in marine species (Rodgers and Stalling, 1972). They showed that the butoxyethanol ester of 2,4-D was hydrolyzed to 2,4-D by fish and that both 2,4-D and the hydrolyzed butoxyethanol ester of 2,4-D were excreted rapidly by fish, but their studies were not designed to characterize metabolites, if any, of the 2,4-D.

In separate experiments, we administered carboxy 14-C labeled phenylacetic acid, 2,4-D, and 2,4,5-trichlorophenoxyacetic acid (2,4,5-T) intravenously (caudal vein) to dogfish sharks (phenylacetic acid, 2,4-D and 2,4,5-T) and flounder (phenylacetic acid). Phenylacetic acid, sodium salt, was injected as an aqueous solution and 2,4-D and 2,4,5-T were administered dissolved in dimethyl sulphoxide. The urinary papillae of the fish were cannulated to facilitate collection of urine. The dogfish shark had balloons attached to their catheters and were allowed to swim freely in tanks during the experiment, while the flounder were kept relatively immobile in mesh cages while urine was collected. 24, 48, and 72 hour urine samples were taken and counted for 14-C to determine what percentage of the dose had been excreted. Samples of urine were also extracted with different solvents and subjected to thin-layer chromatography to determine the chemical form of the 14-C excreted (see James and Bend, 1976, for experimental details). For each compound in each species, more than 90% of the 14-C excreted in urine was a single metabolite of the administered acid. In each case, samples of the metabolite were isolated by preparative thick-layer chromatography using silica gel plates, and the isolated metabolite was recrystallized from methanol. Hydrolysis of each metabolite gave the administered acid and a single ninhydrin positive compound which was identified by paper chromatography and amino acid hydrolysis as taurine. In the case of phenylacetic acid and 2,4-dichlorophenoxyacetic acid, sufficient quantities of the pure metabolites were isolated to obtain infrared spectra. Both compounds had strong absorbances at 1600 cm^{-1} and 1200 cm^{-1}, which are characteristic of the sulphonic group.

In related experiments, dogfish sharks were sacrificed at 4, 24, and 48 hours after doses of 2,4,5-trichlorophenoxyacetic acid (10 mg/kg), and samples of bile collected. Up to 15% of the dose was found in the bile at these times, and in each case the bulk of the 14-C (greater than 95%) was present as the taurine conjugate (Guarino, James, and Bend, unpublished data). Little radioactivity (less than 5%) was found in blood or any of the other organs after 24 hours.

This study demonstrates a qualitative difference in metabolism between fish species and the mammalian species studied. Both classes rapidly excrete phenylacetic acid, 2,4-D and 2,4,5-T, but the fish excreted taurine conjugates of all three carboxylic acids, whereas the mammals excreted 2,4-D and 2,4,5-T largely unchanged, and excreted phenylacetic acid largely as the glycine conjugate.

SUMMARY

The work presented here, and that of others in the field (e.g., Lee et al., 1976), demonstrates that rates of xenobiotic metabolism by oxidative pathways are usually slower in marine than mammalian species, and that lipophilic compounds tend to accumulate in many marine organisms.

The cytochrome P-450-dependent oxidative pathways seem to be qualitatively different in crustacean species as opposed to mammalian and fish species. Although hepatopancreas microsomes from the crustacean species studied contained cytochrome P-450 in amounts higher than found in many fish species, we did not detect any significant oxidative metabolism of the three model substrates by the NADPH-dependent pathway in these microsomes. In addition, hepatopancreas microsomes and cytosol fractions from crustacea inhibitied NADPH-dependent mixed-function oxidative activity in hepatic microsomes from two fish species, suggesting that a factor in hepatopancreas can block a necessary step in the NADPH-dependent oxidation of substrates. However, oxidative metabolism by hepatopancreas microsomes could be measured in the presence of sodium periodate or cumene hydroperoxide.

We studied three pathways of metabolism which are commonly considered as detoxication reactions, namely epoxide hydration and epoxide conjugation with glutathione and amino acid conjugation of certain carboxylic acids. In some species, in vitro activities of the epoxide metabolizing enzymes were as high as those found in mammalian species, and all species studied had epoxide metabolizing activities. Glutathione S-transferase enzymes have many substrates among the chemicals which are classed as xenobiotics, and are probably a valuable defense mechanism in marine species as well as other animals. The three carboxylic acids studied in fish were rapidly excreted as taurine conjugates. The rate of elimination of 2,4-D and 2,4,5-T as conjugates was rapid in the dogfish shark and seemed to be similar to that reported for rats, which do not metabolize 2,4-D or 2,4,5-T to any extent. Phenylacetic acid, as phenylacetyltaurine was excreted rapidly compared with lipophilic compounds, but the half-life of excretion of phenylacetic acid in flounder and dogfish shark was about twice as long as that of phenylacetic acid, excreted as phenylacetylglycine, in rats, and also twice as long as that of phenylacetic acid excreted as a mixture of phenylacetylglycine and phenylacetyltaurine, in ferrets. The lipophilic 2,4,5,2',5'-PCB was also excreted very slowly by two marine species compared with rats. In general, if the same, or similar pathways of biotransformation are utilized by marine and mammalian species, marine species metabolize and excrete xenobiotics much more slowly than the mammalian species.

REFERENCES

Bend, J.R., L.G. Hart, A.M. Guarino, D.P. Rall, and J.R. Fouts. 1976. In *Proceedings of the National Conference of Polychlorinated Biphenyls*. Environmental Protection Agency, Washington, D.C. pp. 292-301.

Bend, J.R., G.E.R. Hook, R.E. Easterling, T.E. Gram and J.R. Fouts. 1972. *J. Pharmacol. Exp. Ther.* 183: 206.

Chasseud, L.G. 1976. In *Glutathione: Metabolism and Function*. Raven Press. pp. 77-114.

Coon, M.J., G.D. Nordblom, R.E. White and D.A. Haugen. 1975. *Biochem. Soc. Transactions* 3: 813-817.

Elmamlouk, T.H., T. Gessner and A.C. Browne. 1974. *Comp. Biochem. Physiol.* 48B: 419-425.

Grunow, W. and C. Bohme. 1974. *Arch. Toxicol.* 32: 217-223.

Hyrcay, E.G., J.A. Gustafson, M. Ingelman-Sundberg and L. Ernster. 1975. *FEBS Letters* 56: 161-165.

James, M.O. and J.R. Bend. 1976. *Xenobiotica* 6: 393-398.

James, M.O., J.R. Bend and J.R. Fouts. 1976. *Biochem. Pharmacol.* 25: 187-193.

James, M.O., R.L. Smith and R.T. Williams. 1972a. *Xenobiotica* 2: 499-506.

James, M.O., R.L. Smith, R.T. Williams and M.M. Reidenburg. 1972b. *Proc. Roy. Soc. Lond.* B. 182: 25-35.

Lee, R.H., C. Ryan and M.L. Neuhauser. 1976. *Marine Biol.* In Press.

Leutz, J. and H. Gelboin. 1975. *Arch. Biochem. and Biophys.* 168: 722-725.

Lowry, O.H., N.J. Rosebrough, A.L. Farr and R.J. Randall. 1951. *J. Biol. Chem.* 193: 265.

Pohl, R.J., J.R. Bend, A.M. Guarino and J.R. Fouts. 1974. *Drug Metab. Disp.* 2: 245-555.

Poland, A.P., E. Glover, J.R. Robinson and D.W. Nebert. 1974. *J. Biol. Chem.* 249: 5599-5606.

METABOLISM OF PESTICIDES BY AQUATIC ANIMALS

M. A. Q. Khan, F. Korte, and J. F. Payne

ABSTRACT

Present state of knowledge of the metabolism of pesticidal chemicals by aquatic vertebrates and invertebrates has been reviewed. *In vivo* and *in vitro* metabolic pathways of organophosphates, organochlorines, carbamates and mercurials have been described. *In vitro* systems that metabolise cyclodienes and their epoxides namely, microsomal mixed-function oxidase, microsomal epoxide hydrase, and soluble glutathime-S-transferase(s) have been investigated in fresh water and marine animals.

INTRODUCTION

In spite of the global concern about the effects of pesticides on foodwebs and foodchains, very little attention has been paid to the fate of pesticides in non-target animals in aquatic environments. As a result very little and insufficient information is available about the metabolic pathway of a single pesticidal chemical in any non-target species. Organized efforts of Brodie and Maickel (1962), Fouts (see: Pohl et al. 1974), Adamson (1974) and Dixon et al. (1967) and their co-workers to characterize *in vitro* the detoxication mechanisms of marine fish especially the drug metabolizing enzymes have provided extensive information about the ability of marine fish to metabolize xenobiotics (Adamson 1974). Studies of the in vitro characterization of similar enzymes in fresh water animals by Dewaide and Henderson (1968, 1970), Buhler and Rasmusson (1968), Murphy (1966), and Khan (see: Stanton and Khan 1973, 1975; Garretto and Khan 1975) and their co-workers have provided information that fresh water fish are capable of degrading pesticidal chemicals and this along with their ability to rapidly eliminate the absorbed pesticides can enable them to survive in toxic environments.

The subject of metabolism of pesticides by biota has been reviewed (O'Brien 1967, Smith 1968, Khan et al. 1974, Menzie 1974, Hathaway 1975, Matsumura 1975). The important metabolic reactions that pesticidal chemicals undergo are: oxidations, hydrolysis and reductions. These may then be followed by the excretion of the primary or secondary metabolites either directly or after conjugation with endogenous molecules (Smith 1968, Adamson 1974, Khan et al. 1975). Esters (organophosphates, carbamates, pyrethroids, phenoxyacetates) are very susceptible to biological hydrolysis while polycyclic aliphatic cyclodienes are metabolized very slowly by living organisms. The enzymes that attack these pesticides include mixed-function oxidase(s) (MFO), epoxide hydrase, glutathione-S-transferase(s), glutathione-dependent dehydrochlorinase, nitro- and azo-reductases and various esterases. This article will provide up-to-date information about in vivo and in vitro metabolism of pesticides by these enzymes of aquatic animals.

Metabolism of Organophosphates

Organic phosphorothioate esters can be attacked at (i) sulfur atom (forming sulfone and sulfoxide or replacing sulfur with oxygen), (ii) ester bonds (by specific esterases) and (iii) alkoxy bonds (forming dealkylated products). In the case of parathion and EPN the nitro-groups can be reduced to amino groups (Fig. 1)

Several fish have been found to metabolize in vivo malathion, parathion, methylparathion (Benke et al. 1974, Ludke et al. 1972), diazinon (Hogan and Knowles 1968), fenitrothion (Gift and Lockhart 1973, Lockhart et al. 1973) and TOCPP (Murrey 1974) although the metabolites have not been characterized in most cases. In vitro metabolism by fish liver enzymes have shown that the MFO can perform the oxidative desulfuration, sulfoxidation and even esteric hydrolysis. Early work using vertebrate liver slices showed that the brook trout and the brown trout could convert parathion to paraxon and deethylated metabolite(s) at rates slower than that by mammalian liver (Potter and O'Brien 1954). In similar comparative studies with liver homogenates Murphy (1966) found variations among fish species in the rates of oxidative desulfuration of parathion, guthion and malathion. Sunfish and bullheads showed rates comparable with those showed by mammalian system (Table 1). However, the rate of hydrolysis of these insecticides by fish liver homogenates was much lower than that by the mammalian preparations (Table 1). The hepatic

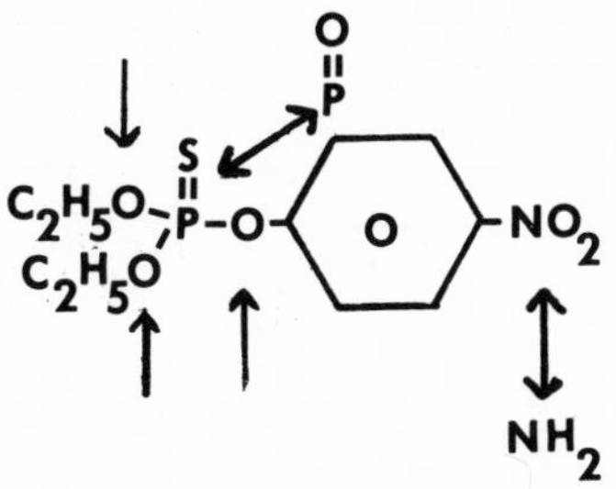

Fig. 1. Metabolic sites (shown by arrows) of parathion.

Table 1. Activation and Hydrolysis of Parathion and Malathion by Liver Homogenate of fish expressed as a percent of the activity of Mouse Preparation

	% of Mouse Activity*			
	Parathion		Malathion	
Animal	Activation	Hydrolysis	Activation	Hydrolysis
Sunfish	260	32	2516	40
Bullhead	189	32	181	24
Flounder	68	11	264	20
Sculpin	5	20	95	32

*Activation and hydrolysis measured by inhibition of acetylcholinesterase by paraoxon and malaoxon; data recalculated from Murphy (1966).

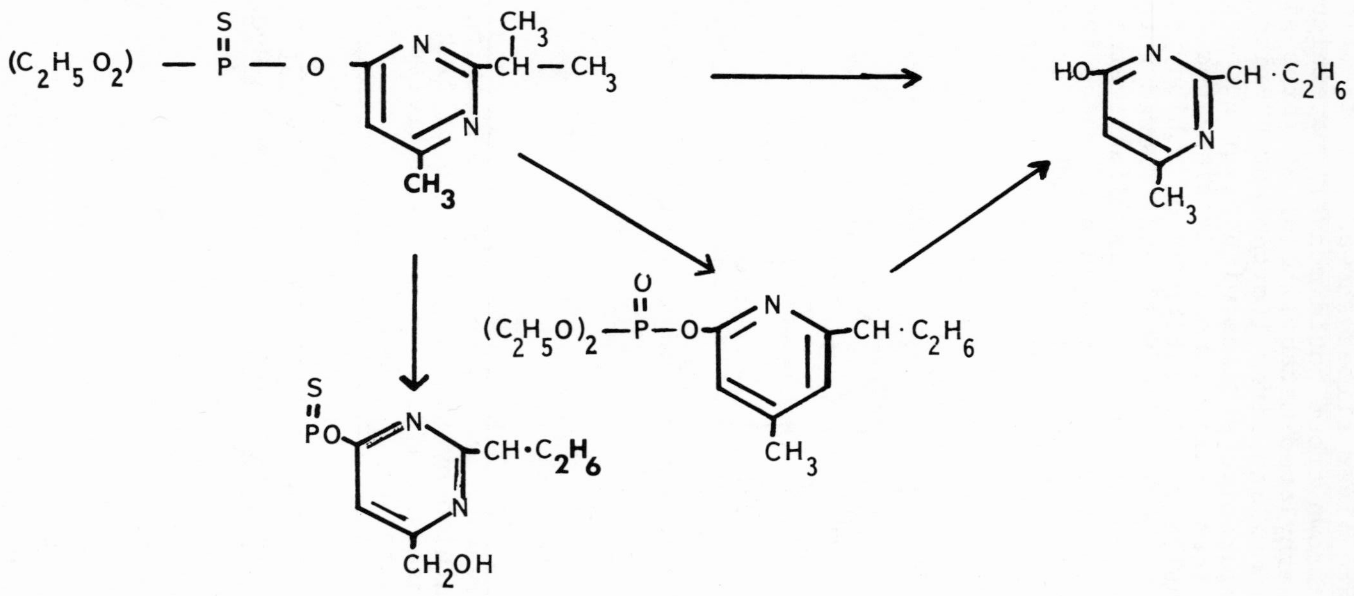

Fig. 2. Metabolic pathways that Diazinon undergoes in aquatic animals (Hogan and Knowles 1968, 1972).

MFO activity as seen by the co-factor requirements in above studies is located in the microsomal fraction. The latter was found in the channel catfish (Ictalurus punctatus) with Diazinon which was metabolized to Diazoxon and to polar metabolites, diethylphosphorothioic acid and diethylphosphoric acid (Fig. 2) (Hogan and Knowles 1972).

The reduction of the nitrogroups to amino groups of parathion and EPN by liver homogenates occurs at almost comparable (flounder, sculpins, largemouth bass, sunfish, and alewife) or lower rates (bullheads, suckers, bluegills) than by mammalian preparations (Table 2)

Table 2. Nitro-reductase Activity of Liver Homogenates (Hitchcock and Murphy 1967)

	μmoles/100 mg/30 min	
Animal	aminoparathion	amino-EPN
Rat	1.256	0.254
Mouse	1.191	.220
Bullhead	1.431	.249
Sucker	1.364	.257
Flounder	.839	.125
Sculpin	.544	.055
Largemouth bass	.913	.145
Sunfish	.954	.155
Bluegill	1.093	.180
Alewife	.850	.136

(Hitchcock and Murphy 1967). This may be an important detoxication reaction in fish which cannot hydrolyze Parathion or Paraoxon or do so at very low rates.

In vivo metabolism of organophosphates by aquatic invertebrates has been studied only in mosquito larvae. Larvae of *Culex pipiens fatigans* can hydrolyze Fenthion to water soluble metabolites: dimethylphosphoric acid, dimethylphosphorothioic acid (most abundant in the body), and an unidentified metabolite. The body residues also included fenoxon sulfone and sulfoxide but the water contained only the sulfoxides of Fenthion and Fenoxon (Stone 1969, Stone and Brown 1969). Larvae of *Aedes aegypti* converted Abate to its sulfoxide, sulfone, oxygen analogue and demethylated derivatives (Fig. 3). Some of these metabolites become conjugated (Leesch and Fukuto 1972). Mosquito larvae can convert Malathion to Malaoxon and its deacylated products (Matsumura and Brown 1963). Several other aquatic invertebrates may be capable of metabolizing Fenitrothion (Lockhart 1973, Lockhart et al. 1974, Zitko and Cunningham 1974).

Unlike fish, the MFO of a few aquatic invertebrates studied cannot metabolize organophosphates. The MFO of the hepatopancreas of the American lobster (*Homarus americanus*) can attack Parathion at a rate 1/8 that of the rat MFO (Carlson 1973) although other studies failed to detect any such *in vitro* metabolism of Parathion and EPN by the hepatopancreas of the quahog (*Mercencria mercenaria*) (Carlson 1972) and of the Lobster (Carlson 1974, Elmamlouk and Gessner 1976, 1976a).

The DFP-ase enzyme present in the homogenate of the optic nerve of the squid can hydrolyze DFP (Deltbarn and Hoskin 1975).

Metabolism of Organochlorines

DDT is metabolized to DDE in most aquatic animals. The residues of DDE in fish following their treatment or that of their habitat with DDT has been reported for goldfish (*Carasus auratus*) (Grzenda et al. 1970; Yang et al. 1971), dogfish (*Squalus acanthias*) (Dvorchik and Maren 1972), mosquitofish (*Gambusia affinis*) (Kapoor et al. 1970; Dziuk and Plapp 1973), Cutthroat trout (Allison et al. 1971), Young Atlantic salmon (Sprague et al. 1970) and other fishes (Bridges et al. 1963; Freedeen et al. 1971, Sprague and Duffy 1971, Reinert et al. 1974, Ernst and Goerk 1974, Juengst and Alexander 1975). DDD has also been reported to be produced from DDT in fish (Grzenda et al. 1970, Kapoor et al. 1970, Yang et al. 1971, Allison et al. 1971). The rate of *in vivo* metabolism of DDT to DDE is much higher in DDT-resistant than in S-strains of the mosquitofish (Dziuk and Plapp 1973).

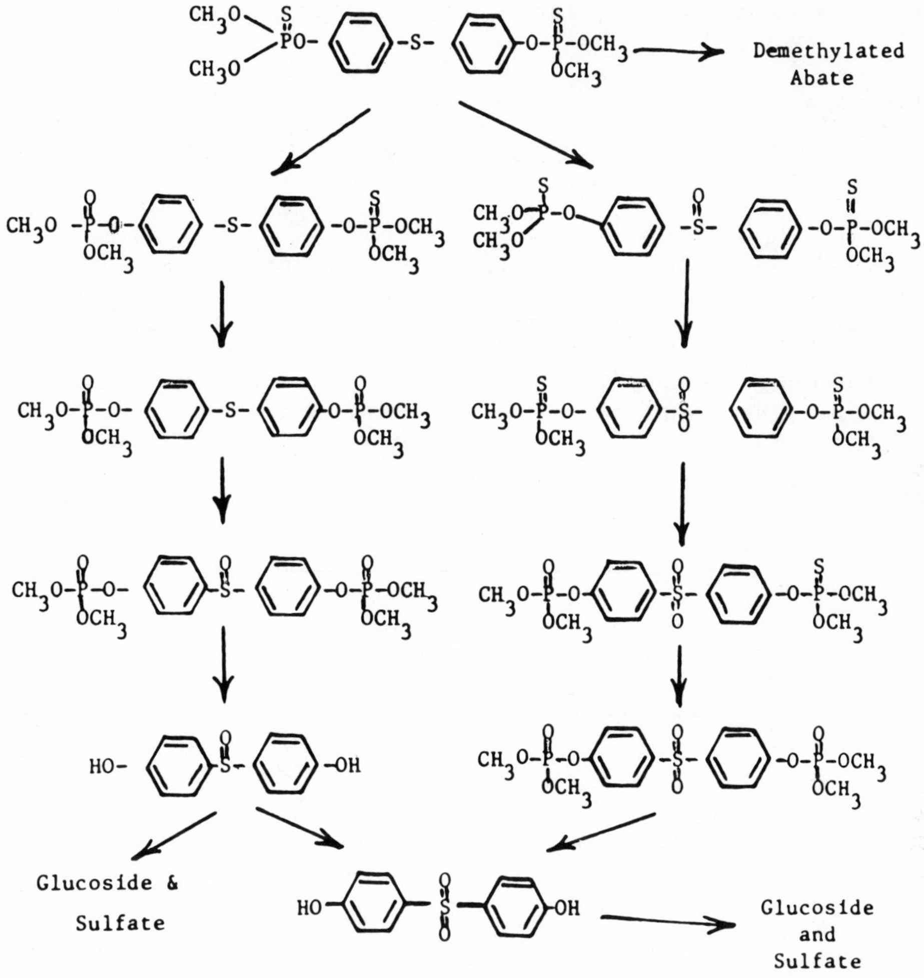

Demethylated Abate

Glucoside & Sulfate

Glucoside and Sulfate

Fig. 3. Metabolism of Abate in aquatic animals (Leesch and Fukuto 1972).

Dehydrochlorination of DDT is also common among aquatic invertebrates. This has been demonstrated *in vivo* and *in vitro* in mosquito larvae (*Aedes aegypti* and *Culex* spp) (Kimura and Brown 1964, Pillai and Brown 1964). Other invertebrates reported to dehydrochlorinate DDT *in vivo* are: leech (*Hirudinea*) (Kimura et al. 1968), annelid worm (*Allolobophora caliginosa*), slug (*Agriolimax reticulatus*) (Davis and French 1969), snail (*Physa*) (Kapoor et al. 1970), crustaceans (*Daphnia* spp.) (Kapoor et al. 1970, Johnson et al. 1971, Neudorf and Khan 1974), slugs (Davison and French 1969, Atchison and Johnson 1975) and other invertebrates (Johnson et al. 1971).

The rate of *in vivo* DDE production varies from 13 to 21% of the total extractable DDT + DDE in *Daphnia pulex* (Neudorf and Khan 1974), *Daphnia magna*, *Gammarus fasciatus* and *Palaemonetes kadiakennis* (Table 3). *Daphnia magna* and *Palaemonetes* also produced about 7% DDD (Johnson et al. 1971) while *D. pulex* did not produce any DDD (Neudorf and Khan 1974). In addition to DDE the metabolite DTMC was present in *Palaemonetes* and *Libellula* (dragonfly niad). *Palaemonetes* also

Table 3. *In vivo* Metabolism of DDT as Measured by the Concentration of Residues Extracted from Freshwater Invertebrates Following Exposure to DDT (Johnson et al. 1971)

Invertebrate	% of Total Body Residue		
	DDT	DDE	DDD
Daphnia magna (Cladoceran)	73.4	19.7	6.6
Gammarus fasciatus (Amphipod)	79.1	20.9	---
Palaemonetes kadiakennis (Decapod)	50.9	13.2	7.2
Hexagenia bilineate	14.9	85.0	
Ischnura verticalis (Odonata)	39.2	60.2	
Libellula sp. (Odonata)	56.3	28.4	
Chironomous sp. (Diptera)	80.8	19.1	

showed the residues of DBP (Johnson et al. 1971).

DDE appears to be more stable in aquatic animals as it can be recovered unmetabolized from the saltmarsh caterpillar (Estigmene acrea) (Metcalf et al. 1975).

Methoxychlor is a biodegradable analogue of DDT. Its half-life in water by the presence of fish can be reduced from 270 to 8 days (Merna et al. 1972). Animals in an aquatic model ecosystem can demethylate, dechlorinate and dehydrochlorinate methoxychlor (Fig. 4). Other metabolites have not been identified (Kapoor et al. 1970).

Methyochlor which is relatively less toxic and more biodegradable than Methoxychlor can be converted by animals in an aquatic model ecosystem to several metabolites, five of which have been characterized (Kapoor et al. 1970). The reactions involve sulfoxidation and sulfone formation (apparently by the MFO) and dehydrochlorination.

Dicamba. In an aquatic model ecosystem, Dicamba and its metabolite, e.g., 5-hydroxy-2-methoxy-3,6-dichlorobenzoate are conjugated or remain in anionic form and can be extracted only after acid hydrolysis (Yu et al. 1975). The acetone extractable residues were: Daphnia NIL, crab 67% of the total radioactivity in the body.

Pentachlorophenol: The shellfish (Tapes philippinarium) can conjugate it as a sulfate ester (Kobayashi et al. 1969, 1970). The protoporphyrin enzyme, peroxidase of snails can oxidize it to pentachlorobiphenyl (Fig. 5).

The bluegill (Lepomis gibbosus) can convert 4(2,4-DB) to 2,4-D a reaction characteristic of microorganisms (Gutenman and Lisk 1965). Marine fish flounder (Pseudopleuronectes americanus), little skate (Raja erinacea) and dogfish (Squalus acathias) differ from mammals in the excretion of phenylacetic acid. This is eliminated slowly via urine and bile. The phenylacetyl-coenzyme-A is readily metabolized by kidney mitochondria in the presence of glutathione (James et al. 1973).

TFM: This commonly used larvicide for lampreys is converted apparently to its glucuronide formed in the bile and in outside water in the rainbow trout. Nitro-reductase converted TFM to aminophenol (*in vitro* which was acetylated by liver and kidney extracts) as well as to its glucuronide in the presence of UDPGA (Lech 1972; Lech and Costrini 1972). Renal excretion of TFM in coho salmon (*O. kisutch*) has also been reported (Hunn and Allen 1975).

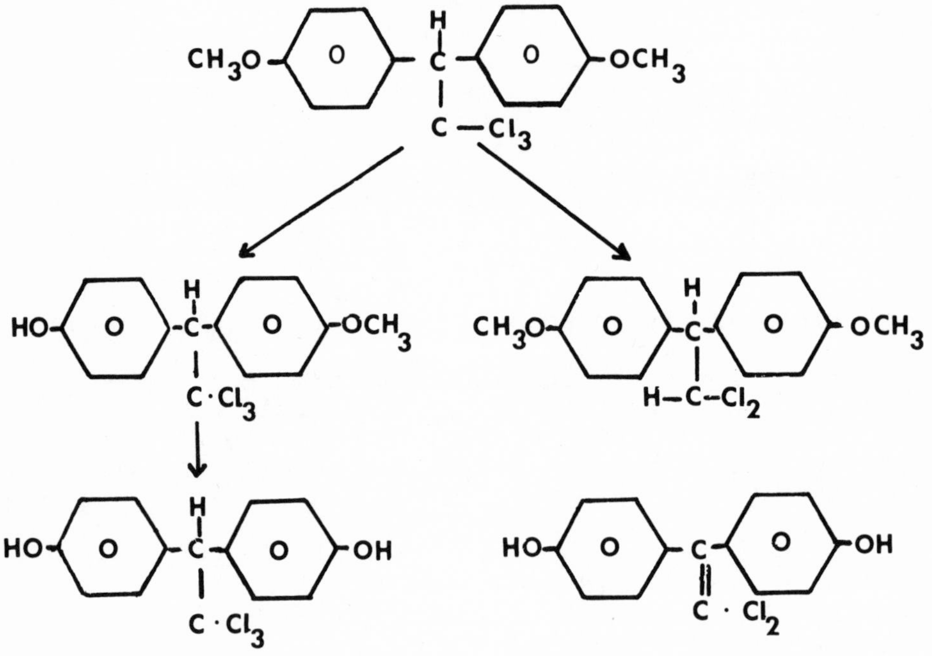

Fig. 4. Metabolism of Methoxychlor in animals of an aquatic model ecosystem (Kapoor et al. 1970).

Bayer 73 (2',5-dichloro-4'-nitrosalicylanilide). Transfer to clean water after a 2-hr exposure of the rainbow trout (*S. gairdneri*) to Bayer 73 gave 10,000:1 ratio of ^{14}C-radioactivity for bile:water in next 12 hours. A glucuronide of Bayer 73 was identified. During next 24 and 48 hours almost all radioactivity disappeared from the tissues (Statham and Lech 1975).

Cyclodienes: Epoxidation of the unsaturated bonds

of cyclodienes occurs commonly in aquatic invertebrates and vertebrates (Khan et al. 1971, Kawatski and Shmulbach 1971, Ludke et al. 1972, Krieger and Lee 1973).

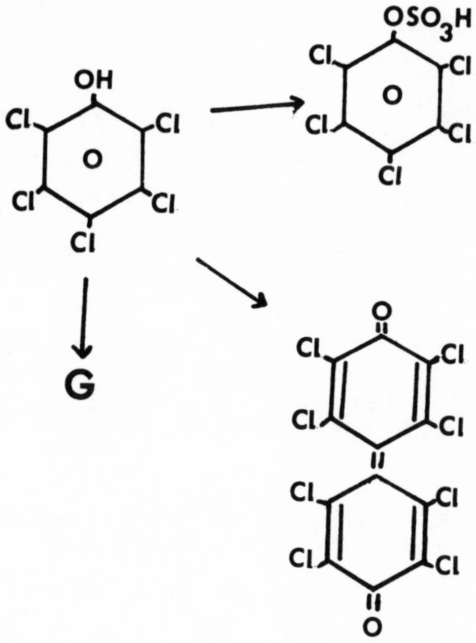

Fig. 5. Metabolism of pentachlorophenol in aquatic invertebrates (Kobayashi et al. 1969, 70). G = glucuronide.

Goldfish exposed to aldrin almost completely epoxidized it to dieldrin in certain tissues (Gackstatter 1968). Aquatic invertebrates from various phyla exposed to aldrin for 2 hours showed the following concentrations of dieldrin (% of total aldrin and dieldrin): <u>Hydra littoralis</u> (Coelentrata, Hydrozoa) 2.59, <u>Dugensia</u> (Platyhelminthes, Plannaria) 1.03, a leech 1.51, <u>Asellus</u> (Arthropoda, Crustacea) 2.30, <u>Gammarus</u> (Arthropoda, Crustacea) 7.07, <u>Daphnia</u> (Arthropoda, Crustacea) 5.69, <u>Cyclops</u> (Arthropoda, Crustacea) 4.13, <u>Cambarus</u> (Arthropoda, Crustacea) 8.48, <u>Aeschna</u> nymph (Arthropoda, Insecta) 24.9, <u>Aedes aegypti</u> larvae (Arthropoda, Insecta) 42.4, <u>Anodonta</u> (Mollusca, Pelecypoda) 19.0, <u>Lymnaea</u> (Mollusca, Gastropoda) 16.8 (Khan et al. 1972). <u>In vivo</u>

epoxidation of aldrin has been reported in the fish golden shiners, mosquitofish, green sunfish, bluegills and catfish (Ludke et al. 1972).

Further metabolism of cyclodiene epoxides by aquatic animals either does not occur or takes place at very low rates. The products are generally monohydroxy and dihydroxy in nature. The latter may follow the hydration of the epoxide. Goldfish eliminated 98% of the absorbed endosulfan within 14 days as diol (Fig. 6) on transfer to insecticide free water. Some fish can conjugate the diol in the bile (Schoethger 1970, Farbwerk 1971, Gorbach and Knauf 1972, Gorbach et al. 1971). Goldfish and bluegills can very slowly metabolize chlordane to lipophilic metabolites (Fig. 7). These metabolites have not been identified (Reddy and Khan, 1977). However Sanborn et al. (1976) were unable to detect any metabolites of chlordanes and toxaphene in extracts of animals in their model ecosystem. Goldfish and Bluegills have been reported to be unable to metabolize dieldrin (Grzenda et al. 1971, Khan and Khan 1974).

Mosquito larvae (*Aedes aegypti*) can hydroxylate aldrin, photoaldrin and telodrin (Korte et al. 1962, Klein et al. 1969, Khan et al. 1973). Dieldrin can be hydroxylated by larvae of *Culex* (Oonithan and Miskus 1964). Mosquito larvae and the crayfish (*Cambarus*) can convert photodieldrin to a liphophilic ketone (Khan et al. 1969, Georgacakis and Khan 1971).

In vitro metabolism of cyclodienes has shed information on various detoxication systems. Epoxidation has been commonly used to measure the level of the MFO activity. The hydroxylation of dihydroanalogues of aldrin and isodrin as well as of chlordene and chlordene epoxide have also been used for MFO characterization. The conversion of the epoxide to yield dihydrodiols is catalyzed by the microsomal epoxide hydrase system. This hydration can also be catalyzed by the soluble glutathione-dependent transferases.

Mixed-function Oxidase: Initial velocities to metabolize various cyclodiene substrates by the hepatic MFO of aquatic animals (Table 4) show marine fish to possess lower levels of activity than the freshwater fish. The latter have lower levels of activity than the mouse (Stanton and Khan 1973, 1975, Garretto and Khan 1975, Khan et al. unpublished data). Kinetic

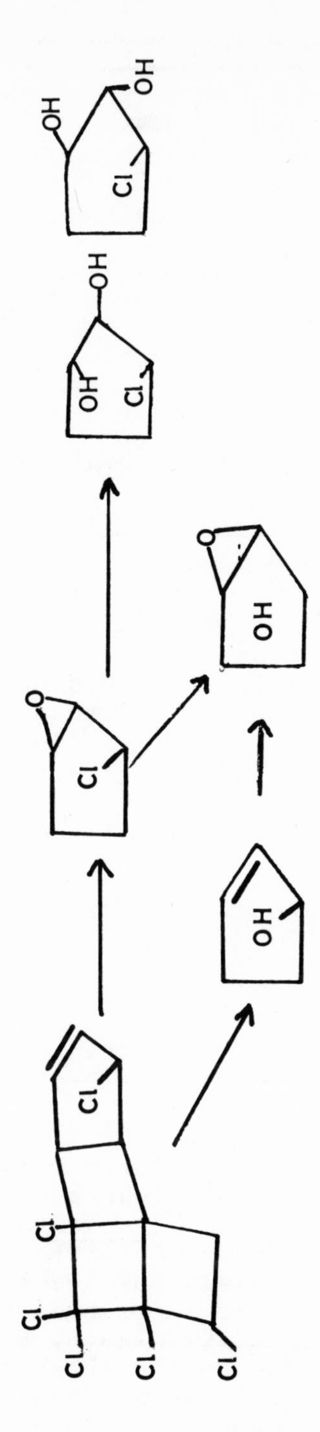

Fig. 6. Metabolism of endosulfan in aquatic animals (Gorbach and Knauf 1972).

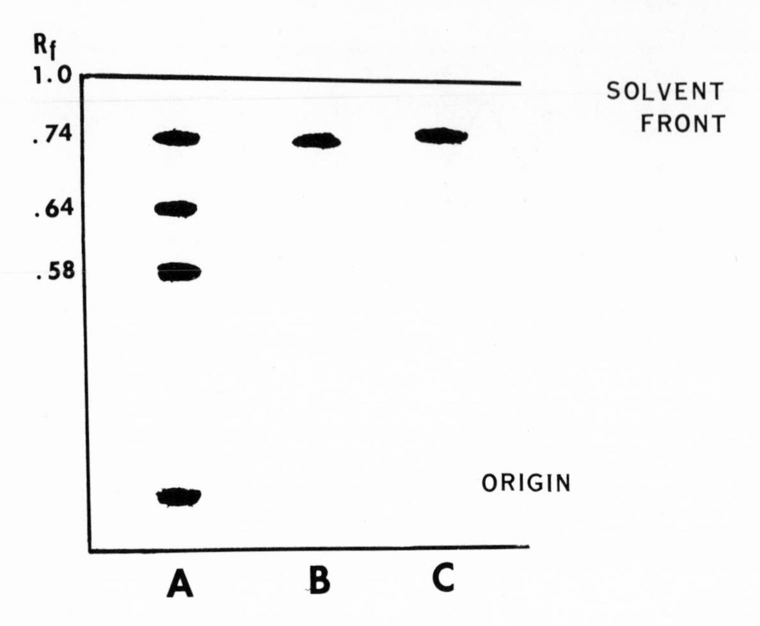

Fig. 7. TLC autoradiograms of extracts of ^{14}C-chlordane treated bluegills (A) and daphnids (B) as compared with authentic chlordane (C) (Reddy and Khan (1977).

constants for the activity of aldrin epoxidase and cytochrome c-reductase and cytochrome P-450 contents are shown in Table 5.

In vitro epoxidation of aldrin by subcell fractions of three freshwater invertebrates, a crayfish (Cambarus), a snail (*Lymnaea*), and a clam (*Anodonta*) showed the activity to reside mostly in the microsomal fraction (Table 6). Similar extremely low level of activity was observed in the hepatic microsomes of the Florida crayfish (*Panilurus argus*) (Tables 5, 6). This low level of activity was due to the absence of the cytochrome c reductase activity even though considerable amounts of P-450 are present. Addition of hydrogen

Table 4. Hepatic MFO Activity Towards Cyclodienes in Freshwater (Stanton and Khan 1973, 1975; Garretto and Khan 1975) and Marine fish and Lobster (Khan et al., unpublished)

	nmole product* formed/min/mg protein							
Fish	D	CE	COH(CEOH)	E	PD	DOH	EOH	CEOH
Bluegill	1.15	,16	1.05	.99	5.16	-	-	-
	.44			.85	-	-	-	-
Bluegill	1.45	.13	.99(NIL)	.81	-	-	-	-
fry	2.45							
Bass fry	1.19	-	-	-	-	-	-	-
Kissing gourami	1.86	.46	.54(.09)	.73	2.50	-	-	-
Oscar	.78	.28	.21(.10)	-	-	-	-	-
Peakcock	.75	.59	.54(.12)	-	-	-	-	-
Barbs	1.05	.85	1.01(.29)	-	-	-	-	-
	.86	.53	.322.13)	-	-	-	-	-
Trout	.61	.55	.13(.11)	-	-	-	-	-
	-	.33	.38(NIL)	-	-	-	-	
Stingray	10	-	-	-	-	2.11	2.45	-
Sheeps-head	.348	-	-	-	-	6.78	1.61	.076
Mullet	.318	-	-	-	-	-	-	-
Flounder	.030	-	-	-	-	-	-	-
Drum	.034	-	-	-	-	-	-	-
Lobster	.010	-	-	-	-	NIL	NIL	NIL
Mouse (male)	3.35	.23	1.13	1.17	2.15	-	-	-

*D = dieldrin for aldrin epoxidation; CE = chlordene epoxide for chlordene; COH = chlordene hydroxide for chlordene; CEOH = chlordene-epoxide hydroxide for chlordene; E = endrin for isodrin epoxidation; PD = photodieldrin for photoaldrin epoxidation; D.OH, E.OH and CEOH in last 3 columns measured by the disappearance of D, E, and CE, respectively.

Table 5. Km and Vmax Values for Aldrin Epoxidation by Freshwater and Marine Fish and Lobster (Stanton and Khan 1975, Garretto and Khan 1975, Khan et al. unpublished)

Animal	Vmax nmole/min/mg	Km μM	P-450 n-mole/ mg	Cyt-c reduc- tase*
Freshwater				
Bluegill				
fry	1.72	6.70; 8.58	-	-
young	.18; 1.28	15.00; 13.00	.81	19.0
adult	.62; .62; 1.44	6.70; 8.58	.81	19.0
Bass fry	1.72	5.97	-	-
Kissing gourami	1.86, .26, .47	15.89; 11.06	1.03	24.0
Trout	.59	-	.56	21.0
Marine Fish				
Sheeps- head	.213	29.5	-	-
Mullet	.151	18.0	-	-
	.088	14.3	-	-
Flounder	.026	32.0	-	-
	.035	58.0	-	-
Stingray	.010	21.0	-	-
Inverte- brate				
Lobster**	10(83)	33(29)	-	-

*n-mole/min per mg protein. ** values in parentheses in the presence of cumene hydroperoxide.

Table 6. Aldrin Epoxidase Activity in the Post-Nuclear-Supernatant (and Microsomes) of the Gut, Liver and Kidney of 3 Freshwater Macroinvertebrates (Khan et al. 1971)

	P-mole dieldrin/mg tissue/hr*		
Organ	Crayfish	Snail	Clam
Gut	4.04 (5.4)	1.28	2.61
Liver	5.48 (3.86, 6.8)	10.13 (9.75)	3.07(2.92)
Kidney	8.86 (8.40)	N.D.	N.D.

*values in parentheses for microsomes

peroxide and/or cumene hydroperoxide did show that the P-450 could epoxidize aldrin at a considerable rate (Table 5). Absence of NADPH-cytochrome c P-450 reductase activity in microsomes of the hepatopancreas of the American lobster (Homarus americanus) may explain the failure to detect MFO activity towards aldrin and parathion and EPN (Elmamlouk and Gessner 1976, Carlson 1972) even though considerable amounts of P-450 were present (Carlson 1974, Elmamlouk and Gessner 1974, 1976a).

Homogenates of the mosquito larvae (Aedes aegypti and Culex pipiens) can hydroxylate aldrin and dieldrin (Korte et al. 1962, Oonithan and Miskus 1964, Klein et al. 1969).

Microsomal Epoxide Hydrase: Hepatic microsomes of several marine fish can hydrate cyclodiene epoxides (Table 7). Freshwater fish show much lower levels of activity towards these epoxides (Khan, unpublished data). The hydration of a dieldrin analogue, HEOM occurred only in the bleak (Alburnes lucida) at extremely low rate (0.029 n-mole hydrated/mg protein per min as compared with 0.45 n-mole/mg/min by male rat) (Walker et al. 1974).

Lobster (Panilurus argus) also show considerable level of epoxide hydrase activity in the microsomal fraction of its hepatopancreas (Table 7).

Table 7. Hepatic Microsomal Epoxide Hydrase Activity Towards Cyclodiene Epoxides in Marine Animals (Khan et al. unpublished)

	n-mole substrate disappeared/min/mg protein			
Substrate*	Stingray	Sheepshead	Mullet	Lobster
Heptachlor epoxide	1.79	1.768;.051	NIL	3.609
Chlordene epoxide	8.86	---	NIL	12.833
Oxychlordane	6.08	NIL	0.896	NIL
Dieldrin	---	2.253	---	2.417
Endrin	3.53	1.729	0.573	2.794

*Incubations (average of 4 replicates) with 0.6 mM substrate and 0.1 mg protein.

Glutathione-S-transferase(s): The two marine fish studied show levels of epoxide-transferase activity in their hepatic cytosol (Table 8). This level is higher than that observed in freshwater fish (Khan, unpublished data).

The products of epoxide metabolism by the above three enzymes have not been identified. Gas chromatographic behavior of the heptachlor epoxide metabolite is similar to that of the dihydroxyheptachlor (Khan et al. unpublished data).

Metabolism of Carbamates

Hydroxylations of the ring and the N-methylgroups and hydrolysis of ester bonds, common reactions of carbamate esters, make these compounds very susceptible to chemical and biological degradation in aquatic environments.

The channel catfish (*Ictalurus punctatus*) rapidly

Table 8. Glutathione-S-Transferase(s) Activity of the Hepatic Cytosol of Marine Fish Towards Cyclodiene Epoxides (Khan et al., unpublished)

Substrate	n-mole Substrate disappeared*/min/mg prot	
	Stingray	Sheepshead
Heptachlor epoxide	0.958	0.58, .218
Chlordene epoxide	-	2.05
Oxychlordane	-	.554
Dieldrin	2.417	1.298
Endrin	2.417	.426, .970

*Average of 4 replicates, incubation with 1 mM substrate and 0.1 mg prot.

eliminate carbaryl (Sevin) and its metabolites on transfer to insecticide-free water (Macek and McAllister 1970). The metabolites in fish may include α-naphthol which is retained in the body (Korn 1973).

The mussel (*Mytilus eludis*) hydrolyze carbaryl to α-naphthol (Armstrong and Meilman 1976). Mosquito larvae (*Culex pipiens fatigans*) can metabolize (*in vitro*) carbaryl to α-naphthol, and 4-hydroxy- and 5-hydroxy-carbaryl; carbofuran to 3-hydroxy- or 3-keto-carbofuran; and aldicarb to sulfoxides, sulfones (Shrivastava et al. 1971) (Table 9). Unidentified organosoluble and water soluble metabolites are also produced. The microsomes prepared from the homogenate of the whole larvae of various mosquito species (*Anopheles albimanus*, *Aedes aegypti* and *triseriatus*, *culex peus*, *tarsalis*, *fatigans*) can metabolize propoxur (1.4 to 2.6 per cent) to 5-hydroxy- , N-hydroxymethyl-, and N-demethylpropoxur, acetone, and unidentified metabolites.

Table 9. Metabolism of Carbaryl, Carbofuran, and Aldicarb by Homogenates of the Larvae of *Culex pipiens fatigans* (Shrivastava et al. 1971)

	% Radioactivity Recovered		
	Carbaryl	Carbofuran	Aldicarb
parent compound	68.2	73.9	61.7
water soluble	14.7	7.7	1.8
organo-soluble*	4.1	8.7	31.9
unknown**	15.1 (5)	9.7 (6)	7.6(4)

*5-hydroxy- and 4-hydroxy-carbaryl; 3-hydroxy- and 3-keto-carbofuran; sulfone, sulfoxide, oxime sulfoxide, nitrile sulfoxide of Aldicarb.
**numbers in parenthesis show the number of metabolites.

Metabolism of Other Pesticides

Rotenone: About 90% of the orally administered (through diet) rotenone was retained by the carp. The metabolites in tissues appeared to be similar to those in insects and rats and mice. The major liphophilic product in the water appeared to be 6',7'-dihydro-6',7'-dihydroxyrotenone II and not 6',7'-dihydro-6',7'-dihydroxyrotenone (Fukami et al. 1970).

The *in vitro* MFO activity (rats and carp) is much higher in the liver than in brain, kidney, or small intestinal. Almost all metabolites were ether soluble (Figure 10). Addition of soluble cytosol to the MFO increased water soluble metabolites in the case of the liver (Table 10). The 1.8-fold rotenone tolerance by a strain of the mosquito fish could be overcome by sesamex indicating the detoxication of rotenone by the MFO (Fabacher and Chambers 1971).

Mercurials: Fish, guppy (*Poecilia reticulata*) exposed in tanks containing mercuric chloride and mercuric sulfide in the sediment showed the maximum methylmercury proportions 30% for metallic mercury (produced by microbial action in sediment), 40% for mer-

Table 10. Metabolism of Rotenone by Rat and Carp Liver Fractions (Fukami et al. 1970)

Enzyme	% Radioactivity as:			
	Rotenone	Rotenone I	ether soluble	water soluble
Rat Liver MFO	3	4	86	7
Rat Liver MFO + Cytosol	1	2	21	76
Rat Liver Cytosol	54	13	8	25
Carp Liver MFO	6	8	60	26
Carp Liver MFO + Cytosol	1	0	25	74

curic sulfide (Gillespie 1971, Gillespie and Scott 1971). The methylation of mercury is due to microorganisms (Jernelov 1969, Jensen and Jernelov 1969). Lignosulfate stimulated methylation of mercury under anaerobic conditions (Gillespie 1971) and livers of tuna and albacore, due to methylcobalamine, methylated mercuric chloride (Imura et al. 1972).

Methylmercury may be produced biologically from dimethylmercury (Jensen and Jernelov 1968). The methylmercury content of fish can vary from 10 to 100% of the total mercury (Ueda and Akoi 1970; Bache et al. 1971). Methylmercury is accumulated in amphibia, mostly in liver (up to 2 ppm) and less in muscle (up to 0.5 ppm) (Bryne et al. 1975). Shellfish also accumulates methyl(methylthio)-mercury (Ueda and Akoi 1970; Bache et al. 1971). Some species-specific metabolic factors and size appear to determine the mercurial levels, but nothing is known about these factors or their influence on mercurial binding (Barber et al. 1972).

Guppy (Lebistes reticulata) and snails (Helisoma campanulata) readily adsorb PMA and convert it to inorganic mercury mainly and to ethylmercuric chloride to a small extent (Fanz 1976).

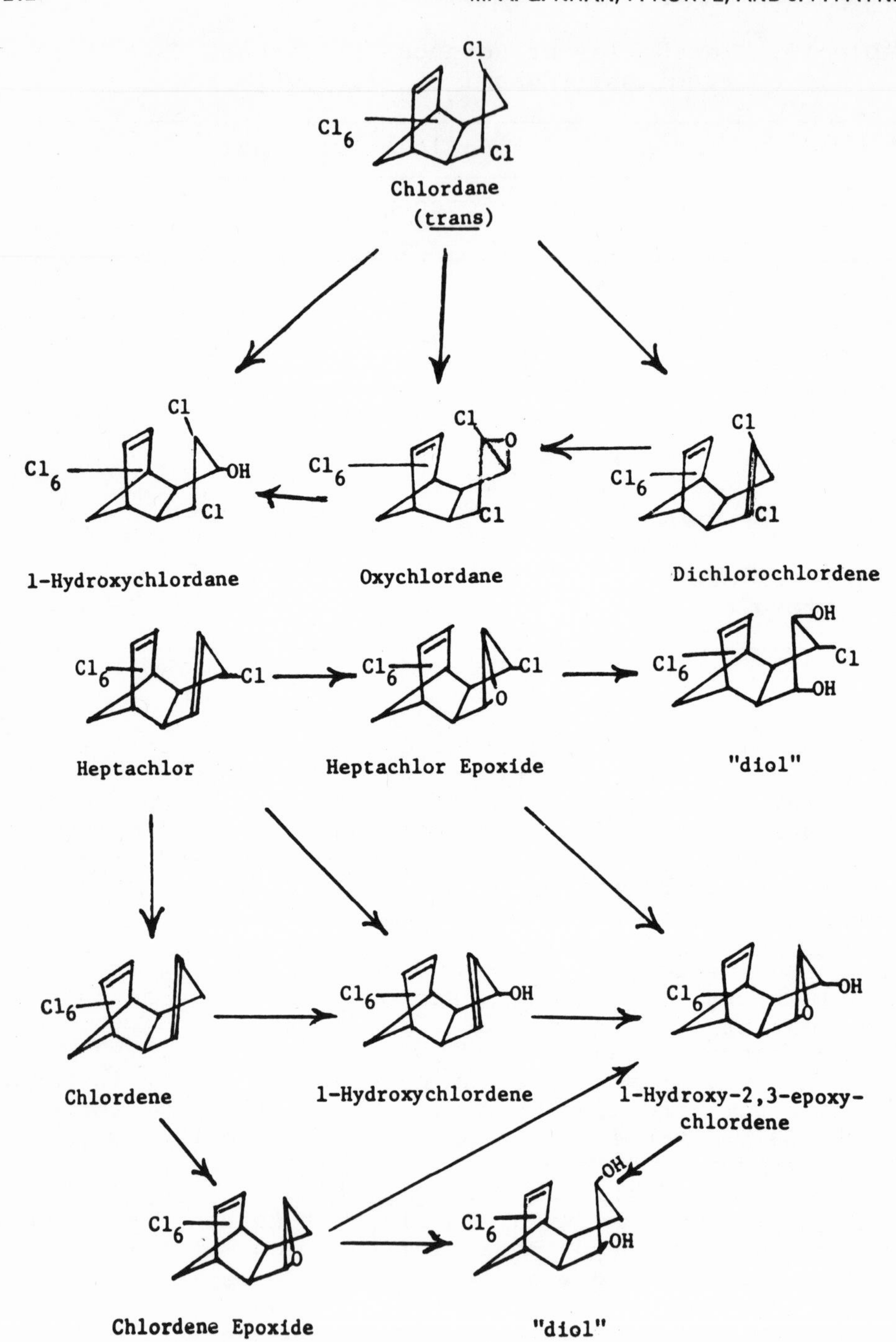

Fig. 8. Metabolic pathways of cyclodienes in aquatic animals (Brooks 1974).

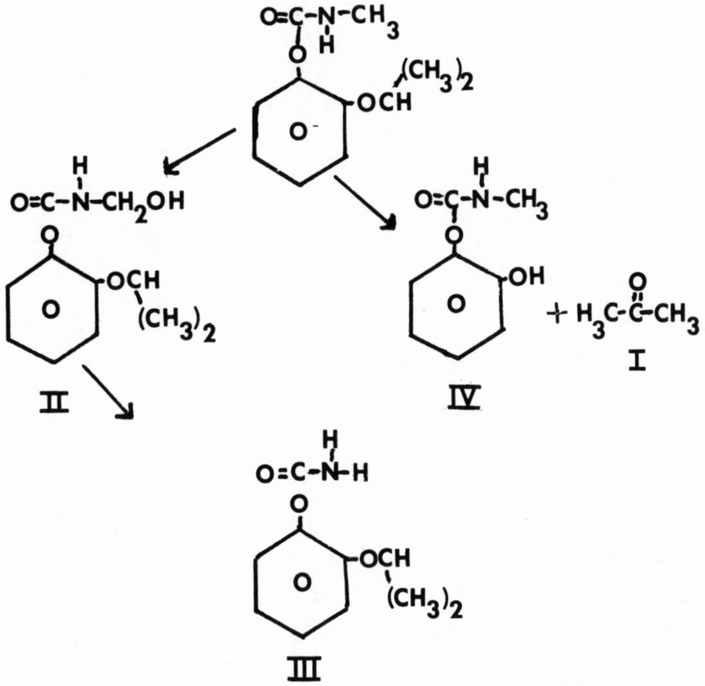

Fig. 9. Metabolism of Propoxur by mosquito larvae (Shrivastava et al. 1970, 1971).

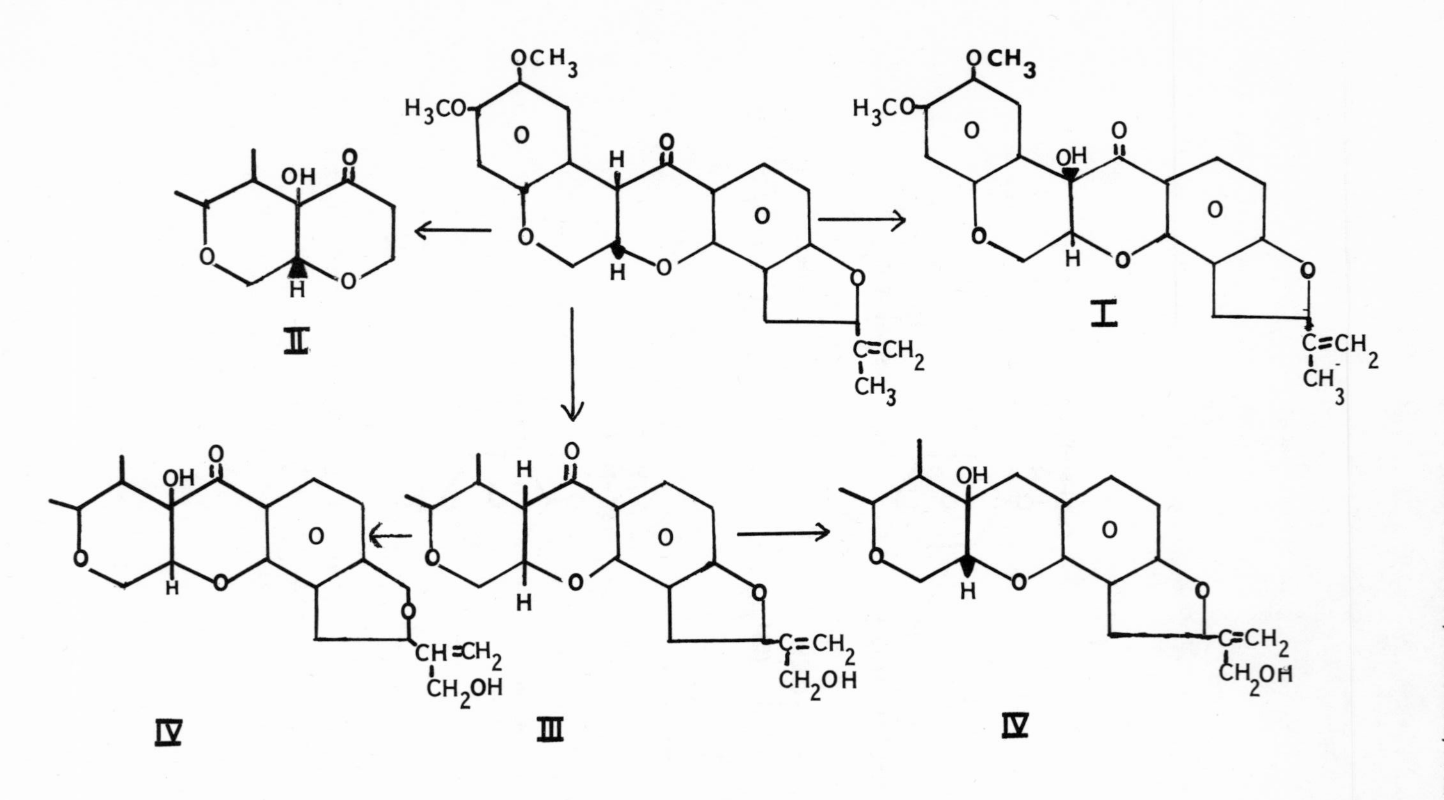

Fig. 10. Metabolism of Rotenone by carp (Fukami et al. 1970).

REFERENCES

R. H. Adamson. 1974. in: Survival in Toxic Environments (M.A.Q. Khan and J.P. Bederka, Jr., eds.). Acad. Press 550 p.

D. Allison, B. J. Kallman, O. B. Cope, and C. C. Van Valin. 1971. Science 143: 958-61.

D. A. Armstrong and R. E. Milemaim. 1976. Marine Biol. 28: 11-16.

G. J. Atchison and H. E. Johnson. 1975. Trans. Amer. Fish. Soc. 104: 775-81.

C. A. Bache, W. H. Gutenmann, and D. J. Lisk. 1971. Science 172: 951-52.

R. T. Barber, A. Vijayakumar, and F. A. Cross. 1972. Science 178: 636-39.

G. M. Benke, K. L. Cheever, F. E. Mirer, and S. D. Murphy. 1974. Toxicol. Appl. Pharmacol. 28: 97-109.

W. R. Bridges, B. J. Kallman, and A. K. Andrews. 1963. Trans. Amer. Fisher. Soc. 92: 421-27.

B. B. Brodie and R. P. Maickel. 1962. in: Proceed. First Internat. Pharmacol. Meeting. Vol. 6. 299-325.

G. T. Brooks. 1974. Chlorinated Insecticides. Vol. II. CRC Press, Ohio 192 p.

A. W. A. Brown. 1963. in: Handbook of Physiology, Vol. V. 773-93. Waverly Publishers, Baltimore, USA.

A. R. Bryne, L. Kosta, and P. Stegnar. 1975. Environ. Letters 8: 147-55.

D. F. Buhler and M. E. Rasmusson. 1968. Compar. Biochem. Physiol. 25: 223-31.

K. A. Burns. 1976. Compar. Biochem. Physiol. 53: 443-46.

G. P. Carlson. 1973. Bull. Environ. Cont. Toxicol. 9: 295-300.

____________. 1972. Compar. Biochem. Physiol. 43: 293-302.

____________. 1974. Bull. Environ. Contam. Toxicol.

D. C. Darrow and R. F. Addison. 1973. Environ. Physiol. Biochem. 3: 196-203.

W. D. Dettborn and F. C. G. Hoskin. 1975. Bull. Environ. Contam. Toxicol. 13: 133-40.

J. H. Dewaide. 1971. Metabolism of Xenobiotics. Comparative and Kinetic Studies as a Basis for Environmental Pharmacology. Drukerj Nijmegen, Netherlands, 164 p.

J. H. Dewaide and P. T. Henderson. 1968. Biochem. Pharmacol. 17: 1901-07.

__________. 1970. Compar. Biochem. Physiol. 32: 489-97.

R. L. Dixon, R. H. Adamson, and D. P. Rall. 1967. in: Sharks, skates, and Rays. (W.P. Gilbert, R. F. Matheson, and D. P. Rall, eds.). John Hopkins Press. 547-552 pp.

L. J. Dziuk and F. W. Plapp. Jr. 1973. Bull. Environ. Contam. Toxicol. 9: 15-19.

W. Ernst and H. Goerke. 1974. Marine Biol. 24: 287-304.

T. H. Elmamlouk and T. Gessner. 1976. Compar. Biochem. Physiol. 53: 19-24.

__________. 1976a. Compar. Biochem. Physiol. 53: 57-62.

__________. 1974. Compar. Biochem. Physiol. 48: 419-25.

D. L. Fabacher and H. Chambers. 1972. Environ. Poll. 3: 139-41.

S. C. Fang. 1976. Archiv. Environ. Cont. Tox. 1: 18-26.

A. G. Farbwerke Hoechst. 1971. Thiodan and the Environment. Technical Bulletin translated by Hoechst, U. K. 1971.

J. Fukami, T. Mitsui, K. Fukunaga and T. Shishido. 1970. in: Biochemical Toxicology of Insecticides (R. D. O'Brien and I. Yamamoto, eds.). Acad. Press. 280 p.

F. J. H. Freedeen, J. G. Saka and L. M. Royer. 197 . J. Fish. Res. Board Can. 28: 105-9.

J. H. Gackstatter. 1968. J. Fish Res. Bd. Can. 25: 1797-1801.

M. Garretto and M. A. Q. Khan. 1975. Gener. Pharmacol. 6: 91-96.

D. C. Gillespie and D. F. Scott. 1971. J. Fish. Res. Board Can. 28: 1807-08.

__________. 1972. J. Fish. Res. Bd. Can. 29: 1035-41.

S. Gorbach, R. Haaring, W. Knauf, and W. J. Werner. 1971. Bull. Env. Cont. Toxicol. 6: 193-97.

__________ and W. Knauf. 1972. in: Environmental Quality and Safety. Vol. I. (F. Coulston and F. Korte, eds.). Acad. Press. 250 p.

N. Graft and W. C. Lockhart. 1974. J. Assoc. Off. Anal. Chem. 57: 1282-1284.

A. R. Grzenda, W. J. Taylor and D. F. Paris. 1972. Trans. Amer. Fish. Soc. 101: 686-90.

__________. 1971. Trans. Amer. Fish. Soc. 100: 215-221.

W. H. Gutenman and D. J. Lisk. 1965. New York Fish and Game J. 12: 108-11.

M. Hitchcock and S. D. Murphy. 1971. Biochem. Pharmacol. 16: 1801-11.

J. W. Hogan and C. O. Knowles. 1972. Bull. Environ. Contam. Toxicol. 1: 61-64.
J. B. Hunn and J. L. Allen. 1975. J. Fish. Res. Bd. Canada 32: 1873-76.
N. Imura, S. K. Pan, and T. Ukita. 1972. Chemosphere 1: 197-201.
B. T. Johnson, C. R. Saunders and H. O. Saunders. 1971. Journ. Fish. Res. Board Canada. 28: 705-9.
F. W. Juengst, Jr. and M. Alexander. 1975. Marine Biol. 33: 1-6.
S. Jensen and A. Jarnellov. 1969. Nature 223: 1453-54.
_____________________. 1968. Nordforst Biocid Information. 14: 3-5.
M. O. James, J. R. Bend and J. R. Fouts. 1973. Mt. Des. Isl. Biol. Bull. 2: 59-62.
I. P. Kapoor, R. L. Metcalf, R. F. Nystrom and G. K. Sangha. 1970. J. Agr. Food Chem. 18: 1145-52.
J. A. Kawatski and J. C. Schmulbach. 1971. J. Econ. Entomol. 61: 316-17.
H. M. Khan and M. A. Q. Khan. 1974. Archiv. Environ. Contam. Toxicol. 2: 289-301.
M. A. Q. Khan, J. R. Bend and J. R. Fouts. (Unpublished). Epoxidation and Epoxide Hydraction of Cyclodienes by Hepatic Subcell Fractions of Marine Animals.
_____________, W. Coello, A. A. Khan, and H. Pinto. 1972. Life Sci. 11: 405-15.
_____________, A. Kamal, R. J. Wolin, and J. Runnels. Bull. Environ. Contam. Toxicol. 8: 219-28.
_____________, J. D. Rosen, and D. J. Sutherland. 1969. Science 164: 318-19.
_____________, E. Toro, R. Moore, and G. Reddy. 1976. in: Pesticides in Aquatic Environments (M. A. Q. Khan, ed.) Plenum Press, 300 p.
_____________, R. H. Stanton, D. J. Sutherland, J. D. Rosen and N. Maitra. 1973. Archiv. Environ. Contam. Toxicol. 1: 159-69.
_____________, R. H. Stanton, and G. Reddy. 1974. in: Survival in Toxic Environments (M. A. Q. Khan and J. P. Bederka, eds.). Acad. Press. 500 p.
_____________, M. L. Gassman and S. H. Ashrafi. 1975. in: Environmental Dynamics of Pesticides (R. Hague and V. H. Freed, Ed.). Plenum Press, 380 p.
_____________. 1977. Unpublished data: In vitro Metabolism of Cyclodiene Epoxides by Freshwater Fishes.
T. Kimura and A. Q. A. Brown. 1964. J. Econ. Entomol. 57: 710-14.

W. Klein, R. Kaul, Z. Parlar, M. Zimmer, and F. Korte. 1969. Beitrage zur Okologischen chemie XIX. Tetrahed Letters 37: 3197-99.
F. Korte, G. Ludwig and J. Vogel. 1962. Liebig Ann. Chem. 656: 135-40.
K. Kobayashi, H. Akitake and T. Tomiyama. 1969. Bull. Jap. Soc. Sci. Fish. 35: 1179-85.
______________________________. 1970. Bull. Jap. Soc. Sci. Fish. 36: 103-109.
R. I. Krieger and P. W. Lee. 1973. Archiv. Environ. Cont. Toxicol. 1: 112-16.
S. Korn. 1973. Trans. Amer. Fish. Soc. 102: 137-38.
J. J. Lech and C. N. Stathan. 1975. Toxicol. Appl. Pharmacol. 27: 300-305.
______ and N. V. Costrini. 1972. Compar. Gener. Pharmacol. 3: 160-66.
R. F. Lee, R. Sauerheber and G. H. Dobbs. 1972. Marine Biol. 17: 201-8.
J. G. Leesch and T. R. Fukuto. 1972. Pest. Biochem. Physiol. 2: 223-35.
W. L. Lockhart, D. A. Metner, and N. Grift. 1973. Mannitoba Entomol. 7: 26-36.
W. L. Lockhart, R. Wagenmann, J. W. Clayborn, G. Graham, and D. Murray. 1975. Environ. Physiol. Biochem. 5: 361-69.
J. L. Ludke, J. R. Gibson and C. I. Lusk. 1972. Toxicol. Appl. Pharmacol. 21: 89-97.
K. J. Macek, C. A. Rodgers, D. Stalling and S. Korn. 1969. Trans. Amer. Fish. Soc. 99: 689-95.
T. H. Maren, L. E. Broder and V. G. Stenger. 1968. Bull. Mt. Des. Isl. Biol. Lab. 8: 39-41.
F. Matsumura. 1975. Toxicology of Insecticides. Plenum Press 600p.
F. Matsumura and A. Q. A. Brown. 1963. J. Econ. Entomol. 56: 381-88.
C. M. Menzie. 1974. Metabolism of Pesticides: An update U.S. Dep. Interior, Fish and Wildlife Series, Special Scientific Report - Wildlife No. 184, Washington, D. C. 485 p.
R. L. Metcalf, J. R. Sanborn, P. Y. Lu, and D. Nye. 1975. Archiv. Environ. Contam. Toxicol. 3: 151-65.
S. D. Murphy. 1966.

D. A. J. Murray. 1975. J. Fish. Res. Bd. Can. 32: 457-60.
G. A. Neville and M. Berlin. 1974. Environ. Res. 7: 75-82.
S. Neudorf and M. A. Q. Khan. 1975. Bull. Environ. Contam. Toxicol. 11: 493-50.

R. D. O'Brien. 1967. Insecticides. Action and Metabolism Acad. Press. 328 p.

N. Pasteur and G. Sinegre. 1975. Biochem. Gen. 13: 789-94.

W. P. Penrose. 1975. J. Fish. Res. Bd. Can. 32: 2385-90.

M. K. K. Pillai and A. Q. A. Brown. 1965. J. Econ. Entomol. 58: 255-66.

J. L. Potter, and R. D. O'Brien. 1954. Science 144: 55-56.

G. Reddy and M. A. Q. Khan. 1977. (Unpublished data).

R. E. Reinert, L. J. Stone and W. A. Willford. 1974. J. Fish. Res. Bd. Can. 31: 1649-52.

J. R. Sanborn, R. L. Metcalf, W. N. Bruce and P. Lu. 1976. Environ. Entomol. 5: 531-36.

D. C. Schmidt and L. J. Weber. 1973. J. Fish. Res. Bd. Can. 30: 1301-09.

R. A. Schoettger. 1970. Toxicology of Thiodan in Several Fish and Aquatic Invertebrates. Investigation in Fish Control. 35. U.S. Dept. Interior, Bureau of Sport Fisheries and Wildlife. U.S. Govt. Print. Office., Washington, D. C.

D. P. Schultz. 1973. J. Agr. Food Chem. 21: 186-92.

S. P. Shrivastava, G. P. Georghion, R. L. Metcalf and T. R. Fukuto. 1970. Bull. World Hlth. Org. 42: 931-92.

S. P. Shrivastava, G. P. Georghion, T. R. Fukuto. 1971. Ent. Exp. Appl. 14: 333-48.

J. N. Smith. 1968. in: Adv. Comp. Physiol. Biochem. 3: 173-232.

J. B. Sprague, P. F. Elson and J. R. Duffy. 1970. Environ. Res. 1: 191-203.

J. B. Sprague and J. R. Duffy. 1971. J. Fish. Res. Bd. Can. 28: 59-64.

C. N. Staltham and J. J. Lech. 1975. J. Fish. Res. Bd. Can. 32: 315-322.

R. H. Stanton and M. A. Q. Khan. 1975. Gener. Pharmacol. 2: 121-27.

________________. 1973. Pest. Biochem. Physiol. 3: 351-57.

B. F. Stone. 1969. J. Econ. Entomol. 62: 977-81.

________ and A. Q. A. Brown. 1969. Bull. World Hlth. Org. 40: 401-08.

T. E. Tooby and F. J. Dubin. 1975. Environ. Bull. 8: 79-80.

R. Wageman, B. Graham and W. L. Lockhart. 1975. Fish. Mar. Serv. Dev. Tech. Res. 486 (Environment Canada).

C. H. Walker, G.A. El-Zegan, A.C. Crane and J.B.R. Kenny. 1974. in: Comparative Studies of Food and Environmental Contamination. Internat. Atom. Energy Agency (Vienna) SM-175/21: 529-40.

ACKNOWLEDGEMENTS

Partial support from a U.S.P.H.S. grant (ES-01479) from N.I.E.H.S. to M.A.Q. Khan is acknowledged. The data on marine animals (Khan, Bend and Fouts, unpublished) were collected while M.A.Q. Khan was associated with the marine pharmacology program of N.I.E.H.S. (1975-76) at the C.V. Whitney Marine Research Laboratory of the University of Florida at Marineland. The latter was made possible by a visiting scientist award by N.I.E.H.S.

DEGRADATION OF DIMILIN® BY AQUATIC FOODWEBS

Gary M. Booth and Duane Ferrell

Abstract

The purpose of this report was to analyze several components of the food web of a lake and pond ecosystem. Specifically, the degradation and accumulation properties of Dimilin (a 25% wettable powder formulation of TH-6040, N-[[(4 chlorophenyl)amino] carbonyl] benzamide) were studied in a model system with water, soil, aquatic vegetation, algae, bacteria, and channel catfish. Following multiple application, Dimilin disappeared from water, sediment, and aquatic vegetation after several days.

Algae (*Plectonema*) degraded 80% of the TH-6040 in a 1 hour incubation period primarily to p-chlorophenyl urea and p-chloroaniline.

Pseudomonas sp. accumulated rather large amounts of TH-6040 from the incubation media when the chemical was used as the sole carbon source; however, no degradation products were detected in the media. Some metabolism occurred when an acetate carbon source was placed in the incubation flasks with the organisms and chemical.

Channel catfish did not bioaccumulate Dimilin residues from treated soil in a simulated lake ecosystem constructed in the laboratory.

Introduction

In living systems energy as well as pesticides are often transferred from one organism to another via the food chain. The term "Food Chain" describes the order or sequence in which one organism preys on another. However, this implied straight-line relationship is usually oversimplified. Most communities have organisms that have alternate food-choices and very often function at several different trophic levels. Thus, the term food web probably best describes the complete ensemble of chains in an entire ecosystem.

Bioaccumulation, movement, and transfer through food webs are important for our understanding of pesticide impact on ecosystems. Accumulation and potential transformation of pesticides in one biological compartment of a web may have a significant effect on the next higher trophic level (Ketchum, 1974).

Aquatic systems are more susceptible to pesticide-food web interactions than terrestrial systems because the water medium maximizes exposure to large numbers of organisms and the systems generally have larger food chains (Craig and Rudd, 1974). Secondary effects of pesticides on aquatic systems such as growth stimulation or inhibition of plankton populations also are important.

When a pesticide enters a water body it may volatilize, remain in the water in solution or suspension as microcrystals or droplets, or absorbed onto particulate matter in the H_2O, or be deposited in soil sediment (Gerakis and Sficas, 1974). Therefore, the principal methods of acquiring pesticide residues by aquatic organisms are direct uptake of water-borne residues and indirect uptake by contaminated food and sediment. The complexity of most food webs is staggering, and one needs to consider the process of pesticide accumulation in its simplest terms. Examination of the behavior of pesticides in water by utilizing organisms from "real" aquatic environments is clearly important.

The intent of this paper is to determine the fate and degradation of Dimilin, a new insect growth regulator, at selected "key" sites in the food web of Utah Lake and the surrounding ponds. Concentration and

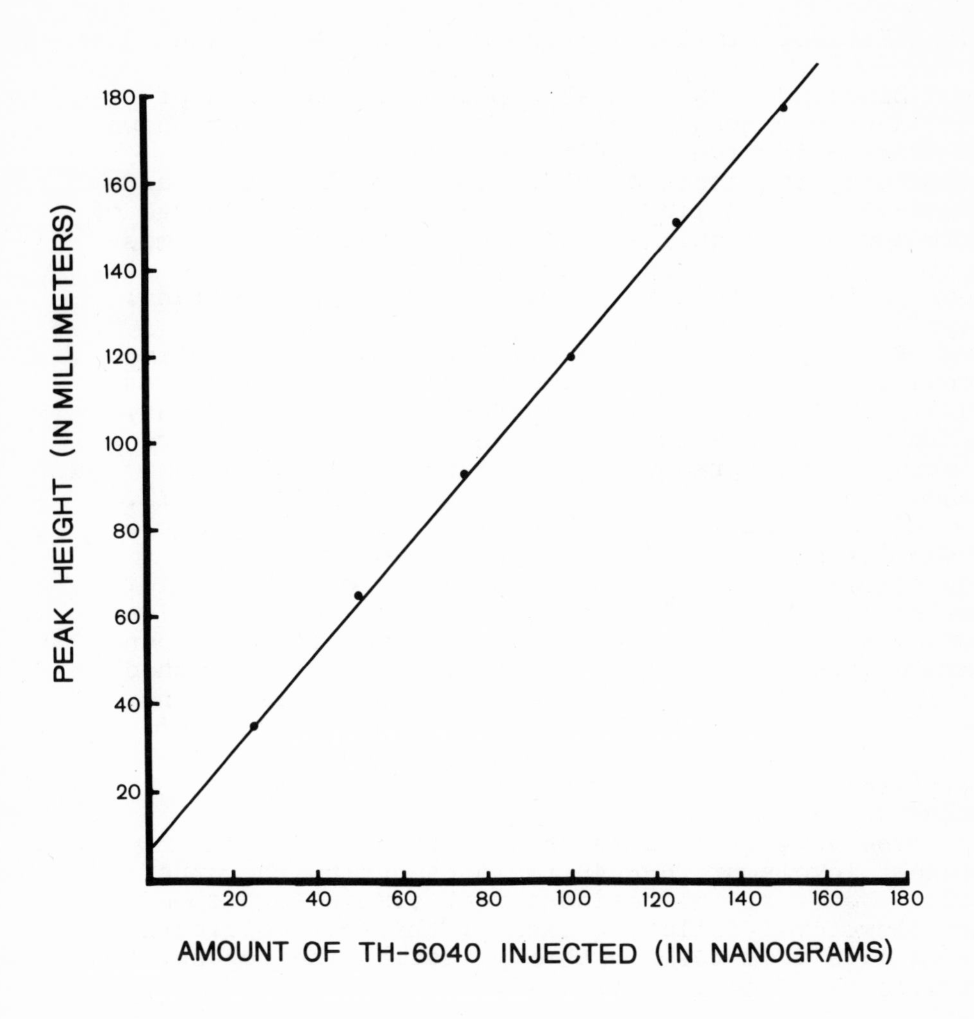

Figure 1. Standard curve for technical TH-6040 using reverse phase high pressure liquid chromatography (HPLC) with an ultra-violet (UV) detection system and a micro-bondapack C_{18} column.

transformation of Dimilin are examined in water, selected microorganisms, soil, algae, and channel catfish that are native to the Utah aquatic ecosystem.

Stability of Dimilin in Pond Water and Non-target Effects

Dimilin, a 25% wettable powder formulation of N-[[(4-chlorophenyl) amino] carbonyl], has been shown to be an effective mosquito control agent at the remarkably low doses of 0.025 and 0.05 lbs. a.i./acre for 8-15 days (Mulla et al., 1974). Since the use of this compound requires multiple applications for mosquito control in aquatic ecosystems, knowledge of its degradation pattern in ponds and lakes is important.

Four applications of Dimilin W-25 were made in ponds located in Salt Lake County between 7/14/75 and 10/7/75. TH-6040 residue was analyzed in the water, sediment and aquatic vegetation using high pressure liquid chromatography with UV sensitivity to 5-10 nanograms (Figure 1). Recoveries of fortified soil, vegetation, and water samples were above 90%. Table 1 shows the results of the pond residue analysis. Detectable TH-6040 residues were not found in any of the soil samples. The residues in the water samples following the first application dropped to background levels (0.009 ppm) after 24 hours, and never reached significant levels even after repeated applications. This water residue data agrees with the results of Schaefer and Dupras (1976) who found TH-6040 to be completely degraded in ponds after 24 hours.

The vegetation data showed consistently the highest levels of TH-6040 residues during the experiment. However, the residues over time also dropped to very low level following each of the four applications. At 28 days from the fourth application the concentration in the vegetation was <0.05 ppm.

These data show that TH-6040 rapidly dissipates and does not accumulate in water, sediment or aquatic vegetation during multiple applications.

Even though TH-6040 disappears from the aquatic environment in just a few days, it was interesting to examine the effects of the chemical on non-target macroinvertebrates (Table 2). Over 20 different species were examined, and only immature corixidae (water boatman) and collembolans (springtails) were significantly effected.

TABLE 1. TH-6040 Water, Soil, and Vegetation Residues Analyzed from Ponds Sprayed Four Times with Dimilin W-25 at 0.04 lbs. A.I./Acre

# APPLICATION	DATE	TIME INTERVAL[a]	SAMPLE RESIDUE (PPM) WATER	SOIL[b]	VEGETATION[b]
	7/14/75	pretreatment	0.004	<0.05	<0.05
First	7/14/75	1 hour	0.036	<0.05	8.2
	7/15/75	1 day	0.009	<0.05	5.2
	7/17/75	3 days	0.007	<0.05	5.4
	7/21/75	7 days	0.016	<0.05	9.3
	7/28/75	14 days	0.006	<0.05	0.84
Second	7/28/75	1 hour	0.019	<0.05	2.7
	8/11/75	14 days	0.003	<0.05	2.0
Third	8/11/75	1 hour	0.005	<0.05	2.0
	8/25/75	14 days	0.001	<0.05	0.51
	9/09/75	28 days	0.005	<0.05	0.24
Fourth	9/09/75	1 hour	0.004	<0.05	4.9
	9/24/75	14 days	0.003	<0.05	0.64
	10/7/75	28 days	<0.001	<0.005	<0.05

[a]Time Interval = lapsed time since each preceding treatment

[b]Limit of detection for soil and vegetation was 0.05 ppm. The soil and vegetation samples were analyzed by Thompson-Hayward Chemical Company.

TABLE 2. Post-Treatment Effects on Non-target Corixids and Collembolans after Multiple Pond Applications of Dimilin W-25 at 0.04 lbs. A.I./Acre. Total Numbers Are Shown. Values in Parentheses Represent Percentages of Corixid and Collembolan Adults and Immatures of the Total Found

	30 DAYS FROM 4TH APPLICATION		80 DAYS FROM 4TH APPLICATION	
SAMPLE	TREATMENT	CONTROL	TREATMENT	CONTROL
Total Adults	75	216	76	84
Total Corixid Adults	1 (1%)	4 (2%)	41 (54%)	61 (73%)
Total Collembolan Adults	0*	151 (70%)	4 (5%)	0
Total Immatures	15	159	12	4
Total Corixid Immatures	5 (33%)*	141 (89%)	0	0
Total Collembolan Immatures	0	0	0	0

*Significant at the 0.05 level of probability.

Table 2 specifically shows that there was a significant difference between the total collembolan adults in the treatment and control 30 days from the 4th application. The collembolan control numbers represented 70% of the total adult organisms examined. Collembolan immatures were not found in any samples.

No differences were observed for the corixid adults in the treatment and control; however, total corixid immatures appeared to be significantly affected 30 days from the 4th application. Corixid immatures represented 33% and 89% of the treatment and control totals respectively.

At 80 days from the 4th application, no significant differences were seen for the treatment and control adult corixids. Hence, depression of corixid immatures is temporary. Large numbers of collembolan adults were not found in either the treatment or control at this later sampling period because of cold weather.

No effects were observed on total population numbers or species diversity.

Depression of non-target populations by Dimilin W-25 has also been reported by Mulla et al. (1976) and Miura and Takahashi (1975) but neither elimination of species nor permanent arrestment of population growth was reported.

Dimilin Effects on Algae

The metabolism of TH-6040 was studied over a 4-day period in a pure culture of a blue-green algae, *Plectonema boryanum*. The initial concentration was 0.1 ppm in the media. The experiment was designed so the total residues could be examined from the algae growth media and the algal cells proper. This was accomplished by filtering the liquid media through a 0.45 micron millipore filter, collecting the algal cells, and weighing the cells on a Cahn Balance with a specially designed micro-pan. Total CO_2 was also trapped over each flask (Table 3).

Plectonema clearly grows rapidly in the presence of TH-6040 with no visible signs of inhibited growth. The algae immediately takes up TH-6040 to a concentration of 144.7 ppm after 1 hour and then rapidly eliminates the residue to a low of 8.3 ppm over the 4 day period.

TABLE 3. Total Residues Expressed as TH-6040 Equivalents in *Plectonema boryanum* (Algae) and Algae Media

DAY	WEIGHT OF ALGAE (MG)		PPM	
	CONTROL	TREATMENT	ALGAE	MEDIA
0	---	---	0	0.1
1 hour	5	5.1	144.7	0.06
1 day	14.1	13.9	85.7	0.06
2 days	21.4	22.5	56.9	0.047
3 days	29.6	30.2	11.7	0.070
4 days	43.4	44.5	8.3	0.076

No $^{14}CO_2$ was trapped above the incubation flasks. A simultaneous drop in the residues occurs in the media after 1 hour incubation followed by an increase in the media residues over time.

This sudden uptake of residues by the algal cells and commensurate drop in the media residues is what one would expect if the organisms were taking up the chemical. Since the residues are slowly eliminated from the algae with an increase in the media residues, it shows a lack of bioaccumulation with time.

Table 4 shows the material balance of the residues in the media. The data shows that after 1 hour, TH-6040 is rapidly metabolized primarily to p-chlorophenyl urea (47.04%) and p-chloroaniline (24.99%) leaving 21.86% of the total as TH-6040. At the end of 4 days, only 4.95% of the total residue was TH-6040, 53.07% was p-chlorophenyl urea, and 35.87% was p-chloroaniline. Trace amounts of activity were found as difluorobenzoic acid and at the origin of the chromatographic plate.

Data from these experiments show that this blue-green algae can rapidly metabolize residues of TH-6040 in water. Since blue-green algae are common constituents of many aquatic food chain species, it follows that residues of pesticides used in aquatic ecosystems could be minimized if they were so designed that they could be easily degraded in this portion of the food web.

It is surprising that just 5 mg of blue-green algal cells can metabolize almost 80% of the TH-6040 present in the liquid media in just 1 hour and 45 mg of the cells can metabolize 95% of the TH-6040 at the end of 4 days. In other words, 5 mg of algal cells can degrade 8 micrograms of TH-6040 in the first 1 hour of incubation.

Dimilin and *Pseudomonas* sp.

Pseudomonas sp. a ubiquitous organism isolated directly from the soil was obtained from the aquatic environment of Utah Lake. A series of 500 ml flasks were set up on a laboratory shaker with 100 ml of growth media in each one. Two types of growth media were used for *Pseudomonas*. Flask A contained all of the necessary salts for growth plus acetate as a carbon source. Flask B contained all of the necessary salts for growth but no carbon source. These flasks and media are hereafter referred to as flask A and flask B for discussion purposes.

TABLE 4. Material Balance of the Total Residues in *Plectonema* Media Analyzed by Thin-Layer Chromatography

Day	TLC spot no.[a]	%	ppm
1 hour	1	4.00 ± 0.25	0.0025 ± 0.0002
	2	47.04 ± 4.45	0.0282 ± 0.0016
	3	1.67 ± 0.06	0.0010 ± 0
	4	24.99 ± 7.63	0.0165 ± 0.0012
	5	21.86 ± 7.42	0.0152 ± 0.0056
1 day	1	6.67 ± 1.33	0.0033 ± 0.0014
	2	52.02 ± 11.10	0.0311 ± 0.0001
	3	3.27 ± 0.63	0.0018 ± 0.0003
	4	28.70 ± 13.41	0.0147 ± 0.0036
	5	8.45 ± 3.38	0.0049 ± 0.0003
2 days	1	5.33 ± 1.35	0.0014 ± 0.0006
	2	51.45 ± 1.61	0.0235 ± 0.0033
	3	4.27 ± 0.78	0.0019 ± 0.0003
	4	36.36 ± 3.31	0.0169 ± 0.0042
	5	2.39 ± 0.43	0.0016 ± 0.0001
3 days	1	5.00 ± 2.48	0.0040 ± 0.0026
	2	45.92 ± 7.82	0.0380 ± 0.0133
	3	3.17 ± 0.45	0.0022 ± 0.0003
	4	32.89 ± 3.59	0.0221 ± 0.0048
	5	10.26 ± 9,83	0.0049 ± 0.0043
4 days	1	4.45 ± 1.98	0.0035 ± 0.0016
	2	53.07 ± 6.55	0.0403 ± 0.0031
	3	2.89 ± 1.89	0.0022 ± 0.0013
	4	35.87 ± 0.21	0.0273 ± 0.0037
	5	4.95 ± 1.59	0.0037 ± 0.0012

[a] 1, origin; 2, p-chlorophenyl urea; 3, difluorobenzoic acid; 4, p-chloroaniline; 5, TH-6040.

TABLE 5. Total Residue (ppm as TH-6040 Equivalents) Observed in *Pseudomonas* and Media of the Acetate-Containing Flasks (Flask A) and Non-Acetate Flasks (Flask B)

DAY	FLASK A (CARBON SOURCE) Pseudomonas PPM	FLASK A (CARBON SOURCE) MEDIA PPM	FLASK B (NO CARBON SOURCE) Pseudomonas PPM	FLASK B (NO CARBON SOURCE) MEDIA PPM
0	0	0.0996	0	0.1000
2	61.7	0.0052	9,519	0.0231
4	62.7	0.0021	10,240	0.0306
6	75.0	0.0051	6,610	0.0275
8	56.9	0.0021	3,278	0.0118
10	61.1	0.0028	5,454	0.0085
12	64.0	0.0059	8,313	0.0384

A known amount of the bacterium was placed into each of the series of flasks and a total of 0.1 ppm of ^{14}C TH-6040 was incubated with each flask. The $^{14}CO_2$ was trapped by blowing compressed filtered air over each flask from a manifold system. Every 48 hours a given set of flasks A and B were analyzed for degradation products. The media was filtered through a 0.45 micron millipore filter to collect the total bacteria for weight determination.

Table 5 shows the total residues in the media and Pseudomonas over the 12 day period. In flask A (acetate as a carbon source) the total ^{14}C activity in the media decreased from 0.0996 ppm to 0.0052 ppm in two days. This left approximately 6% of the original radioactivity in the media. Flask A also shows that on the 2nd day there was an increase in the total residues in Pseudomonas from 0 to 61.7 ppm. This residue level remained fairly constant in Pseudomonas of flask A.

In flask B (no carbon source) the residues in the media dropped from 0.1 ppm to 0.0231 ppm on the 2nd day. The residue remaining in the media after two days represented about 23% of the original activity. However, the Pseudomonas organisms of flask B contained 9,519 ppm on the 2nd day. This high level of radioactivity in Pseudomonas might be explained by the fact that 1) there was 240 fold less bacteria in flask B or 2) TH-6040 was the sole carbon source in the media and the organisms were possibly incorporating ^{14}C TH-6040 carbon atoms into their cellular chemistry. This high level of ^{14}C remained in the flask B Pseudomonas organisms throughout the 12 day experiment.

Pseudomonas in flask A grew very rapidly in the first 24 hours to 1.24×10^9 bacteria/ml, and remained at this population level throughout the study. In flask B, the bacteria population reached a level of 5.2×10^6 bacteria/ml, and this level remained constant in the media over the 12 day study. While Pseudomonas can use TH-6040 as a sole carbon source, they do not grow well compared to the acetate media.

Table 6 shows the material balance of the media residues from both flasks. In flask A, >90% of the ^{14}C was TH-6040 at all sampling periods except days 6, 8, and 10. Day 10 was particularly interesting since 77% of the remaining residues was TH-6040 and 11% was p-chlorophenyl urea. From these data it can be

TABLE 6. $\bar{X}$ Material Balance of Total *Pseudomonas* Media Extractable Residues Expressed in Percent and ppm. The Data Were Obtained by Thin-Layer Chromatography

DAY	FLASK	TLC SPOT NO. *	%	PPM
0	A	1	0.309	1.48×10^{-4}
		2	0.286	1.37×10^{-4}
		3	0.386	1.85×10^{-4}
		4	6.25	2.99×10^{-3}
		5	92.77	0.0444
	B	1	0.247	7.38×10^{-5}
		2	2.502	1.51×10^{-4}
		3	0.110	3.28×10^{-5}
		4	1.34	6.376×10^{-3}
		5	95.80	0.0229
2	A	1	0.466	9.61×10^{-5}
		2	0.705	1.455×10^{-4}
		3	0.552	1.139×10^{-4}
		4	1.197	2.47×10^{-4}
		5	95.045	.01962
	B	1	0.062	1.82×10^{-5}
		2	0.3155	9.25×10^{-5}
		3	0.183	5.4×0^{-5}
		4	1.07	3.14×10^{-4}
		5	97.99	0.0287
4**	B	1	0.177	3.88×10^{-5}
		2	0.748	2.496×10^{-4}
		3	0.299	9.97×10^{-5}
		4	3.133	1.043×10^{-3}
		5	94.801	0.0316
6**	A	1	2.223	9.15×10^{-5}
		2	5.89	2.426×10^{-4}
		3	2.762	1.138×10^{-4}
		4	2.706	2.76×10^{-4}
		5	82.46	0.0034

TABLE 6 (Continued)

DAY	FLASK	TLC SPOT NO. *	%	PPM
8	A	1	0	0
		2	11.2	$3x10^{-4}$
		3	3.275	$9x10^{-5}$
		4	3.26	$9x10^{-5}$
		5	80.975	$2.15x10^{-3}$
	B	1	0.348	$4x10^{-5}$
		2	1.7	$1.9x10^{-4}$
		3	0.911	$1x10^{-4}$
		4	1.25	$1.4x10^{-4}$
		5	94.775	0.01036
10	A	1	1.586	$4x10^{-5}$
		2	11.0625	$2.9x10^{-4}$
		3	5.49	$1.4x10^{-4}$
		4	4.82	$1.2x10^{-4}$
		5	77.32	$1.98x10^{-3}$
	B	1	0.199	$2x10^{-5}$
		2	1.77	$1.8x10^{-4}$
		3	0.4095	$4x10^{-5}$
		4	2.373	$2.4x10^{-4}$
		5	96.737	$9.32x10^{-3}$
12	A	1	0.392	$2x10^{-5}$
		2	5.268	$3.03x10^{-4}$
		3	0	0
		4	3.094	$1.78x10^{-4}$
		5	93.2735	$5.37x10^{-3}$
	B	1	0.155	$7x10^{-5}$
		2	0.7135	$3.2x10^{-4}$
		3	0.126	$6x10^{-5}$
		4	1.695	$7.7x10^{-4}$
		5	94.815	$4.28x10^{-2}$

*1 = origin; 2 = p-chlorophenyl urea; 3 = difuorobenzoic acid; 4 = p-chloroaniline; 5 = TH-6040.

** Flask A and B were missing from days 4 and 6 respectively.

concluded that metabolism or degradation of TH-6040 occurs very slowly in the presence of alternative carbon sources. No metabolism was detected in flask B media. When the organisms are forced to use TH-6040 as a carbon source, they apparently retain the chemical in their cellular structure.

No $^{14}CO_2$ was detectable in the traps of our experiments. Metcalf et al. (1975) exposed Dimilin for 6 hours to Pseudomonas putida and found no evidence of degradation upon extraction. Microorganisms in polluted water have also shown to have little effect on Dimilin (Schaefer and Dupras, 1976).

Degradation of Dimilin by Catfish

Channel catfish (Ictalurus punctatus) are at the top of several food chains in the Utah Lake aquatic food web. In addition, they are bottom feeders and could also pick up pesticide residues from bottom sediments.

A series of 30 gallon aquarium tanks were set up to determine the uptake, metabolism, and bioaccumulation of Dimilin from aged Dimilin-treated soil. A ^{14}C-Dimilin W-25 preparation was thoroughly incorporated into 1814g samples of soil containing 34% sand, 42% silt, and 24% clay. Two concentrations each of 0.55 ppm and 0.007 ppm were studied in replicated experiments. The soil samples were layered into the separate aquaria and a field moisture capacity of 70% was incorporated into each sample. This constituted an aerobic system, and it was sampled over a 14 day period. At the end of the aerobic portion, a 1 inch layer of water was placed over the soil. Water and soil were then sampled for 14 days. This part of the experiment constituted the anaerobic portion. The tanks were then filled (20 gallons), 90 channel catfish were placed into each tank and soil, water, and catfish were sampled for 28 days (uptake phase).

Tables 7 and 8 show the results of combustion analysis of the soil and water samples for the aerobic, anaerobic and uptake periods of the 0.55 and 0.007 ppm tanks respectively. Both types of tanks showed that the residues are apparently released from the soil very slowly over time. In the 0.55 ppm tanks the soil released about 64% of the initial 0.55 ppm soil residues. However, only 2.3% of the residues could be accounted for in the water at the end of the uptake

TABLE 7. Total Residues (ppm) in Soil and Water of Aerobic and Anaerobic Aging Periods and Uptake Period of 0.55 ppm Tanks. The Percent Activity in the Water Is Also Shown

PERIOD-DAY		SOIL PPM	H_2O PPM	% ACTIVITY IN WATER
Aerobic -	0	.55	--	--
	1	.575	--	--
	2	.593	--	--
	3	.566	--	--
	7	.497	--	--
	14	.593	--	--
Anaerobic	1	.371	.0315	5.7
	2	.382	.0052	.95
	3	.433	.0065	1.2
	7	.3161	.0083	1.5
	14	.502	.0092	1.7
Uptake	1	.588	.011	2.0
	3	.422	.0094	1.7
	7	.398	.013	2.4
	10	.366	.011	2.0
	14	.462	.0056	1.0
	28	.217	.0127	2.3

TABLE 8. Total Residues (ppm) in Soil and Water of Aerobic and Anaerobic Periods and the Soil of the Uptake Period of the 0.01 ppm (Actual ppm = 0.0071) Tanks. The Percent Activity in the Water Is Also Shown

PERIOD-DAY		SOIL PPM	H_2O PPM	% ACTIVITY IN WATER
Aerobic -	0	.0071	--	--
	1	.00586	--	--
	2	.0073	--	--
	3	.0051	--	--
	7	.0068	--	--
	14	.0067	--	--
Anaerobic	1	.00465	0	0
	2	.0062	0	0
	3	.0063	0	0
	7	.0038	0	0
	14	.0058	0	0
Uptake	21	.00352	0	0
	24	.0045	0	0
	28	.00297	0	0

TABLE 9. $\bar{X}$ Material Balance of Diethyl Ether Extractable Residues from the Anaerobic Water of the 0.5 ppm Tanks

PERIOD	DAY	TLC SPOTS*	%	PPM
Anaerobic	7	1	0.3	0.00002
		2	95.6	0.0079
		3	2.5	0.00021
		4	1.2	0.00010
		5	0.4	0.00003
	14	1	0.6	0.00006
		2	93.4	0.0086
		3	4.5	0.00041
		4	0	0
		5	1.5	0.00014

*1 = origin; 2 = p-chlorophenyl urea; 3 = difluorobenzoic acid; 4 = p-chloroaniline; 5 = TH-6040.

TABLE 10. $\bar{X}$ Material Balance of Methanol Extractable Residues from the 0.5 ppm Soil

PERIOD	DAY	TLC SPOTS*	%	PPM
Aerobic	7	1	0.4	0.0020
		2	15.0	0.0750
		3	0	0
		4	10.9	0.0545
		5	73.7	0.3685
	14	1	0.3	0.00178
		2	7.1	0.0421
		3	0.1	0.0006
		4	16.9	0.1002
		5	75.5	0.4477
Anaerobic	7	1	0.3	0.00095
		2	8.2	0.08914
		3	0.3	0.00095
		4	10.7	0.0155
		5	80.5	0.2545
	14	1	0.6	0.00301
		2	7.5	0.0377
		3	0.2	0.0010
		4	10.7	0.0537
		5	81.2	0.4076
Uptake	14	1	1.4	0.0065
		2	7.1	0.0328
		3	0.2	0.0009
		4	9.9	0.0457
		5	82.7	0.3821
	28	1	0	0
		2	4.4	0.0096
		3	0	0
		4	12.1	0.0263
		5	83.5	0.1812

*1 = origin; 2 = p-chlorophenyl urea; 3 = difluorobenzoic acid; 4 = p-chloroaniline; 5 = TH-6040.

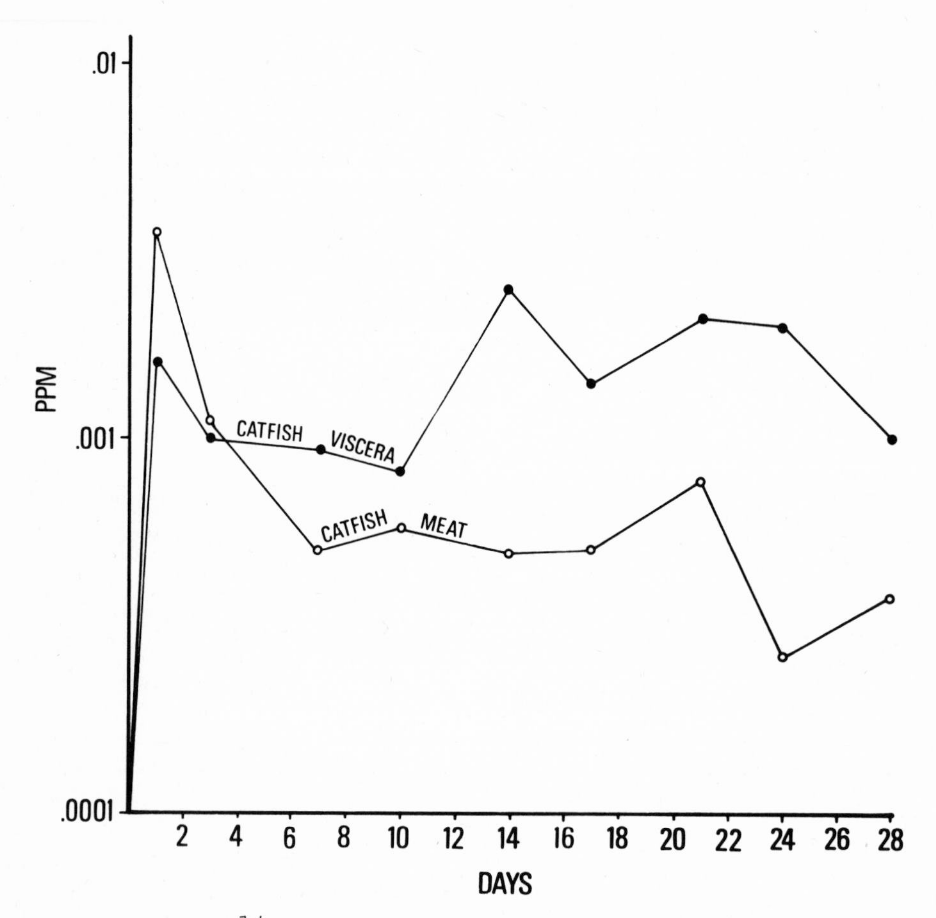

Figure 2. Total ^{14}C residues in catfish "meat" and "viscera" during the uptake period for the 0.01 ppm tanks.

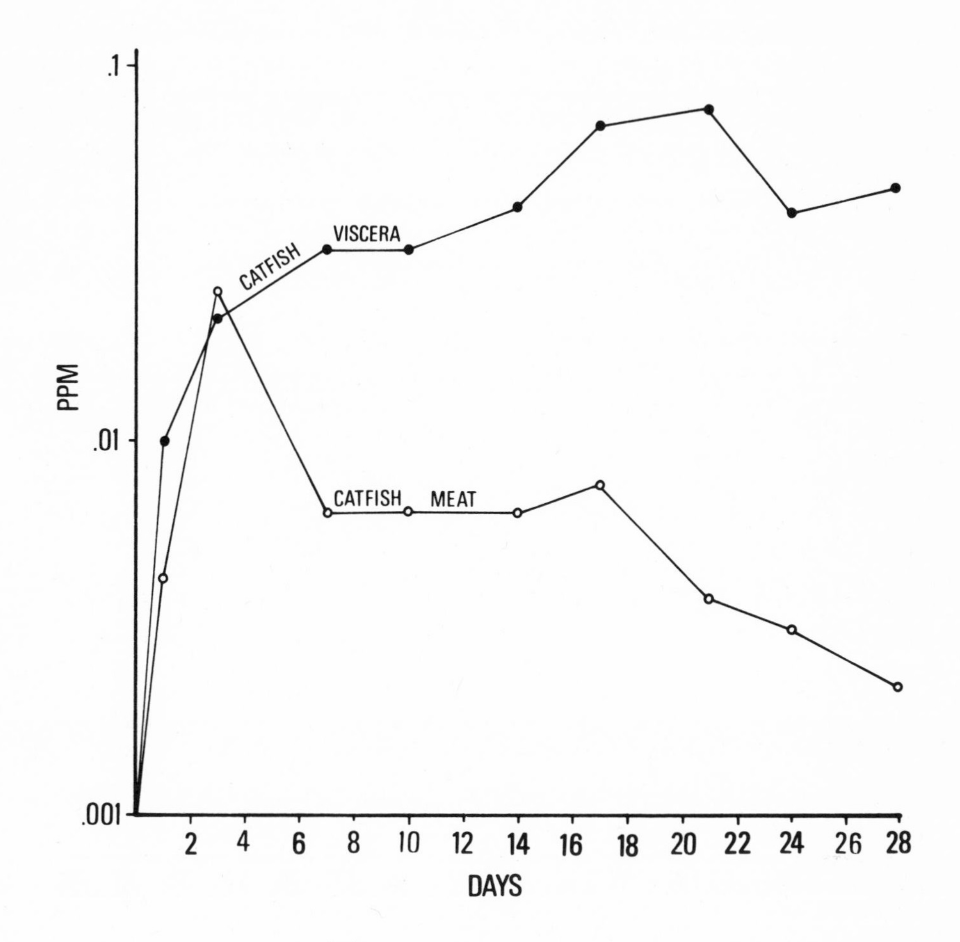

Figure 3. Total ^{14}C residues in catfish "meat" and "viscera" during the uptake period for the 0.5 ppm tanks.

period. The .007 ppm soil samples released about 57% of the initial soil residues into the water, but the residues in the water were too low to be detected accurately.

The water residues from the anaerobic period of the 0.55 ppm tanks were extracted and analyzed for a material balance (Table 9). It is apparent that >93% of the released soil residues was p-chlorophenyl urea. This agrees with the results obtained from field applications of Dimilin W-25 where it has been shown that p-chlorophenyl urea is the major water metabolite from pond treatments (Schaefer and Dupras, 1976).

Analysis of the extractable soil residues is shown in Table 10. An average of 65.9% of the soil residues were extractable leaving approximately 34% as the total bound residues. At the end of the uptake period, it appears that 83.5% of the extractable residues was TH-6040 and 12% was p-chloroaniline. Trace amounts of p-chlorophenyl urea were still found in the soil.

Figures 2 and 3 show the results of the residues found in "meat" (fish minus gut, head and tail) and "viscera" (mainly gut). On the first day in the 0.007 ppm tanks, the "meat" and "viscera" residues rose to about 4 and 2 ppb respectively and plateaued for the remainder of the uptake phase.

In the 0.55 ppm tanks the "meat" and "viscera" residues rose to 4 and 10 ppb respectively and then oscillated up and down over the 28 day uptake period. The concentration of residues seemed to plateau at 3 days for the "meat" and 14 days for the "viscera." At the end of the 28 day uptake-period, the "meat" contained about 2.2 ppb and the "viscera" contained 48 ppb. The residues in tissue were too low for an accurate analysis of the material balance.

These data clearly show that TH-6040 does not bioaccumulate in channel catfish under the conditions of this experiment. Since under marsh or pond application conditions, Dimilin soil residues were shown to be <0.05 ppm, it is not likely that catfish would be contaminated during multiple mosquito application conditions.

References

Craig, R.B. and R.L. Rudd. 1974. In: "Survival in Toxic Environments." M.A.Q. Khan and John P. Bederka, Jr. (eds.) p. 1-24. Academic Press, New York.

Gerakis, P.A. and A.G. Sficas. 1974. In: "Pesticide Review." F.A. Gunther and J.D. Gunther (eds.) p. 69-88. Springer-Verlag, New York.

Ketchum, B.H. 1974. In: "Ecological Toxicology Research." A.D. McIntyre and C.F. Mills (eds.) p. 285-300. Plenum Press, New York.

Metcalf, R.L., P.Y. Lu, and S. Bowlus. 1975. _J. Ag. Fd. Chem._ 23(3): 359-364.

Miura, T. and M.M. Takahashi. 1975. _Mosq. News._ 35 (2): 154-159.

Mulla, M.S., G. Majori, and H.A. Darwazeh. 1975. _Mosq. News._ 35(2): 211-216.

Schaefer, C.H. and E.F. Dupras, Jr. 1976. _J. Ag. Fd. Chem._ Submitted manuscript.

INDEX